## AMONG OUR BOOKS

### Publications in English

- Wave and Wind Directionality. Applications to the Design of Structures
- Symposium on the Pressuremeter and its Marine Applications
- Seabed Reconnaissance and Offshore Soil Mechanics for the Installation of Petroleum Structures. P. LE TIRANT.
- The Use of Anchors in Offshore Petroleum Operations. A. PUECH.
- Glossary of Onshore and Offshore Pipelines. English-French/French-English
- Dictionary of Seismic Prospecting. English-French/French-English.

### Publications in French

- Vagues et ouvrages pétroliers en mer. G. SUSBIELLES et C. BRATU.
- Reconnaissance des sols en mer pour l'implantation des ouvrages pétroliers. P. LE TIRANT.
- La technique des ancres dans l'exploitation pétrolière en mer. A. PUECH.
- Forage. Contrats et statut de l'engin maritime. J. LIGONIE.
- Le positionnement en mer.
- La plongée profonde
- Connaissance de la houle, du vent, du courant pour le calcul des ouvrages pétroliers.
- Climatologie de la mer. Sea Climatology.
- Pénétration sous-marine. Underwater Operations and Techniques.
- Lexique des pipelines à terre et en mer. Anglais-Français/Français-Anglais.
- Dictionnaire de prospection sismique. Anglais-Français/Français-Anglais.
- Guides pratiques pour les ouvrages en mer.
  1. Assemblages tubulaires soudés. ARSEM.
  2. Ancres et lignes d'ancrage. ARGEMA.
  3. Capacité portante des pieux. ARGEMA.

# DESIGN GUIDES
# FOR OFFSHORE STRUCTURES

# welded tubular joints

## ARSEM
Association de Recherche sur
les Structures Métalliques Marines

Translated from the French by Nissim Marshall

Technical advisor: I. Ryan, Centre Technique Industriel
de la Construction Métallique, on technical matters

1987

ÉDITIONS TECHNIP 27 RUE GINOUX 75737 PARIS CEDEX 15    techniP

Translation of
« Guides pratiques sur les ouvrages en mer.
Assemblages tubulaires soudés. »
Association de Recherche sur les Structures
Métalliques Marines (ARSEM).
© Editions Technip, Paris 1985

© 1987 Editions Technip, Paris

ISBN 2-7108-0530-8

Printed in France
by Imprimerie Chirat, 42540 Saint-Just-la-Pendue

# Foreword

Editions Technip have started the publication of a new collection with the general title:

"Design Guides for Offshore Structures"

a title which accurately characterizes the field of application concerning the design, construction, installation and use of the various types of offshore structure built in particular for the petroleum industry.

These Design Guides are being prepared by the offshore structure research associations presented below, whose research and development programs are coordinated under the Offshore Petroleum Research Committee (CEPM, Comité d'Etudes Pétrolières Marines) and with the financial backing of the French Government.

In the past twenty years, the petroleum and allied industries have considerably broadened their field of activity to include the seas and oceans of the entire world, giving birth to a new body of discipline, calling on the competences of marine engineering as well as civil engineering. The characteristics of the new structures designed to support miscellaneous specific equipment of petroleum production, whether fixed or floating, or steel or concrete, for easy or difficult seas, entailed an improvement, indeed a basic change in traditional concepts.

The evolution of the engineering sciences and technologies has been deeply marked by the new problems posed by the action of wave, wind and current on structures installed offshore, at increasing distances from the coast, in ever-deeper waters, and in increasingly severe weather and oceanographic conditions.

On the initiative of Institut Français du Pétrole (IFP) and Institut Français de Recherche pour l'Exploitation de la Mer (IFREMER), four research associations were set up jointly by IFP and IFREMER, already mentioned, the oil companies Elf Aquitaine and Total Compagnie Française des Pétroles, along with contractors and consulting engineers in the petroleum equipment and services industry, and various research institutions and laboratories such as CEBTP, CTICM and Bureau Veritas:

- Association de Recherche sur l'Action des Eléments Marins (ARAE) was formed in 1970. Its activity is the study of environmental parameters, wave, winds and currents, and their action on the structures, in order to determine their seaworthiness.

- Association de Recherche en Géothechnique Marine (ARGEMA) was formed in 1977. It deals with the dimensioning and behavior of the foundations and anchorages of offshore structures.

- Association de Recherche sur le Béton en Mer (ARBEM) was formed in 1978. Its function is to improve knowledge about the behavior of offshore concrete structures, for better prediction and a safer guarantee of the permanence and safety of these kinds of structure.

- Association de Recherche sur les Structures Métalliques Marines (ARSEM) was formed in 1983, taking over the work of a Fatigue Group formed in 1976. It is concerned with the behavior of steel structures at sea, particularly in relation to fatigue mechanisms.

A liaison committee (CLAROM, Conseil de Liaison des Associations de Recherche sur les Ouvrages en Mer) was formed in 1982 to harmonize the overall effort of these associations and to promote the development of their activities in France and abroad. A publications committee was set up within this framework, grouping the Project Managers of the four associations, whose publications should have a decisive impact on developments in engineering techniques.

Each of the Design Guides in preparation concerns a specific subject. Based on state-of-art knowledge, they offer consulting engineers and technicians, builders and operators, knowhow, and recommendations for improving the safety of the structures, while cutting costs.

These recommendations are not meant to serve as regulations, which must in any event be established by qualified organizations, the classification societies, national regulations and international conventions.

However, the regulatory provisions could refer to the Design Guides in so far as the latter incorporate the very latest advances, as, for example, in a new safety assessment method or in a commentary stipulating the range of validity of a given formula.

As further advances are achieved, these Design Guides will have to be revised, and will accordingly be enriched by their confrontation with reality.

We wish the new collection of Design Guides for Offshore Structures to be useful above all to the engineers and technicians faced with the many problems raised by the exploration and production of subsea hydrocarbon fields, and, in general, to all those concerned with the design of offshore structures.

We also wish these Design Guides, prepared by each of the research associations, in its own field, to be developed in close cooperation in accordance with the common objective of structural safety, so as to favor the elaboration of a uniform body of doctrine.

And finally, we wish to ensure that the efforts conducted in the past to expand knowledge in the areas of meteorology, oceanography, hydrodynamics, strength of materials, geotechnics, etc., will contribute to the success of future achievements thanks to our Design Guides, and that the research will continue, with even greater intensity, to meet not only the present needs of the industry, but also future developments.

Pierre Willm<br>
Scientific Director of Marine Engineering at<br>
Institut Français du Pétrole<br>
Chairman of the Publications Committee on<br>
"Design Guides for Offshore Structures"

# Preface

From the practical standpoint, this Guide represents the culmination of the research program on fatigue in welded steel offshore structures initiated in late 1976 by a group formed by Centre National pour l'Exploitation des Océans (CNEXO), Institut Français du Pétrole (IFP) and Société Nationale Elf Aquitaine (Production) (SNEA-P), assisted by a Fatigue Technical Committee, and continued, since 1983, by Association de Recherche sur les Structures Métalliques Marines (ARSEM).

These researches are conducted as part of a French national research program coordinated by Comité d'Etudes Pétrolières Marines. The first phase of this work, which extended from 1977 to 1979, was intended to improve knowledge about the fatigue behavior of welded tubular joints subjected to random loads due to natural elements (wave, wind, current). It comprises two complementary parts, one theoretical and the second experimental:

(1) the theoretical part was entrusted to Laboratoire de Mécanique des Solides at the Ecole Polytechnique (LMS).
(2) the experimental part was entrusted to Institut de Recherches de la Sidérurgie (IRSID).

The experimental part represented the French contribution to a vast European research program on fatigue in welded offshore structures, co-funded by the Commission of European Communities (ECSC) and the member countries participating in the program: United Kingdom, Federal Republic of Germany, Netherlands, Denmark and Italy. Norway joined the program subsequently.

The results of this high quality work were presented at two International Conferences, one held in Cambridge (United Kingdom) from 27 to 29 November 1978, and the second in Paris (France) from 5 to 8 October 1981.

The second phase of this project, which was conducted from 1980 to 1982, and was based on theoretical and experimental knowledge gained during the first phase, was intended to give the designers, builders and users of welded steel offshore structures practical recommendations concerning the fatigue design of these structures, obviously corresponding to the state of the art at the time.

The present Design Guide, whose preparation was entrusted to the Centre Technique Industriel de la Construction Métallique (CTICM) with the assistance of five specialized working groups which included members of LMS, IRSID and the Fatigue Technical Committee (now ARSEM), offers a summary of design methods and knowledge about the design and analysis of welded tubular joints for the construction of offshore petroleum structures built of steel.

This objective led its authors to go beyond the strict framework of the problems raised within the "European Program on Fatigue in Welded Offshore Structures", to deal with other problems likely to influence the dimensioning of a joint, or to affect the fatigue strength of a welded joint.

To derive a better overall view of the parameters or factors likely to affect the fatigue behavior of the structures, it was decided to devote a number of sections to:

(a) corrosion protection, with a general review of the techniques, problems and effects on fatigue strength.

(b) steel grade selection methods and welding, to highlight specific construction problems, especially concerning those encountered in welding very thick materials.

(c) miscellaneous approaches in structural analysis, to pinpoint the hypotheses and, above all, to emphasize their "deterministic", "probabilistic" or "random" character.

(d) static strength of tubular joints, because, in the chronology of engineering design calculations, these rules are the prerequisite to joint dimensioning.

Certain provisions, rules and approaches recommended in this Design Guide may be different from and may even disagree with those specified in a given set of official regulations. Hence it does not claim to replace any particular specification, but rather to reflect the present state of the art, with its range of certainties, and evidently its complementary range of uncertainties, within which it was necessary to take the final decisions.

The essential purpose of this Design Guide is to inform. Although it has been compiled according to the rules of the art, with all the necessary care and attention, based on scientifically checked data, the information it contains. cannot be used unless the conditions of its practical application for a specific project have been duly interpreted by a qualified engineer. The publication of this Design Guide is not a guarantee on the part of ARSEM or of any other natural or legal person, mentioned among the authors, of its relevance to any general particular application, or an encouragement to waive any regulation in force.

ARSEM research projects are initiated by the Institut Français du
Pétrole (IFP) with the assistance of the Institut Français de Recherche
pour l'Exploitation de la Mer (IFREMER) and conducted chiefly by LMS,
IRSID, Centre Technique Industriel de la Construction Métallique
(CTICM) and Bureau Veritas.

The composition of ARSEM was as follows on 1 January 1985:

- Bouygues Offshore (BOS),
- Bureau Veritas (BV),
- Centre Technique Industriel de la Construction Métallique
  (CTICM),
- Compagnie Française d'Entreprises Métalliques (CFEM),
- Compagnie Française des Pétroles (CFP),
- Entrepose GTM pour les travaux Pétroliers Maritimes (ETPM),
- Institut Français du Pétrole (IFP),
- Institut Français de Recherche pour l'Exploitation de la Mer
  (IFREMER)
- Institut de Recherches de la Sidérurgie Française (IRSID),
- Laboratoire de Mécanique des Solides (LMS), Ecole Polytechnique,
- Sambre et Meuse (plants and steelworks),
- Société Française d'Etudes d'Installations Sidérurgiques (SOFRESID),
- Société Nationale Elf Aquitaine (Production) (SNEA-P),
- Union Sidérurgique du Nord et de l'Est de la France (USINOR).

# Acknowledgements

This Design Guide, whose preparation was entrusted to the Centre Technique Industriel de la Construction Métallique, is based on the discussions and recommendations issued within the following five specialized working groups.

## GT1    Forces and loads

| | |
|---|---|
| Mr. Willm (leader) | IFP |
| Mr. Barnouin | IFREMER |
| Mr. Brucker | SAFETEC |
| Mr. Chabrolin | CTICM |
| Mr. Deleuil | C.G. Doris |
| Mr. Dumas | C.G. Doris |
| Mr. Falcimaigne | IFP |
| Mr. Gauvrit | SOFRESID |
| Mr. Goyet | CTICM |
| Mr. Jaunet | Bureau Veritas |
| Mr. Lempire | SAFETEC |
| Mr. Lemeur | SOFRESID |
| Mr. Planeix | Bureau Veritas |
| Mr. Renard | CTICM |
| Mr. Susbielles | IFP |
| Mr. Thébault | SNEA(P) |

## GT2    Stress concentrations

| | |
|---|---|
| Mr. Godeau (leader) | SNEA(P) |
| Mr. Brozzetti | CTICM |
| Mr. Bury | Bureau Veritas |
| Mr. Cabiran | SNEA(P) |
| Mr. Gascouin | SOFRESID |
| Mr. Gérald | SNEA(P) |
| Mr. Legras | ETPM |
| Mr. Mézière | LMS, Ecole Polytechnique |
| Mr. Radenkovic | LMS, Ecole Polytechnique |
| Mr. Recho | CTICM |
| Mr. Ryan | CTICM |

## GT3   S-N curves

| Mr. Brozzetti (leader) | CTICM |
| Mr. Andréau | Bureau Veritas |
| Mr. Bastenaire | IRSID |
| Mr. Cabiran | SNEA(P) |
| Mr. Foucriat | SOFRESID |
| Mr. Gérald | SNEA(P) |
| Mr. Goyet | CTICM |
| Mr. Huther | Bureau Veritas |
| Mr. Lieurade | IRSID |
| Mr. Putot | IFP |
| Mr Recho | CTICM |
| Mr. Ryan | CTICM |

## GT4   Cumulative damage

| Mr. Huther (leader) | Bureau Veritas |
| Mr. Andréau | Bureau Veritas |
| Mr. Bignonet | IRSID |
| Mr. Brozzetti | CTICM |
| Mr. Chabrolin | CTICM |
| Mr. Chauchot | IFREMER |
| Mr. Dang-Van | LMS, Ecole Polytechnique |
| Mr. Foucriat | SOFRESID |
| Mr. Gérald | SNEA(P) |
| Mr. Goyet | CTICM |
| Mr. Lemoyne | IFREMER |
| Mr. Radenkovic | LMS, Ecole Polytechnique |
| Mr. Recho | CTICM |
| Mr. Truchon | Creusot-Loire |

## GT5   Steels for platform tubular structures

This Working Group was formed within the Naval Technical Committee, Offshore Commission of Bureau Veritas.

| Mr. Brozzetti (leader) | CTICM |
| Mr. Charleux (secretary) | Bureau Veritas |
| Mr. Berlin CSFT | CSFT |
| Mr. Bourges | Creusot-Loire |
| Mr. Calinaud | UIE |
| Mr. Chaussy | Forges et Aciéries de Dilling |
| Mr. Debiez | IS |
| Mr. Devillers | IRSID |

| | |
|---|---|
| Mr. Divry | Vallourec |
| Mr. Giraud | CNIM |
| Mr. Guillaud | Chantiers de l'Atlantique |
| Mr. Huard | B.G. Engineering, Bouygues Offshore |
| Mr. Juglar | CFEM |
| Mr. Kieffer | Vallourec |
| Mr. Lempire | SAFETEC |
| Mr. Liégeois | IS |
| Mr. Lieurade | IRSID |
| Mr. Metz | CFP-Total |
| Mr. Monchaud | SNEA(P) |
| Mr. Mouty | COMETUBE |
| Mr. Rousseau | CSS/BNS |
| Mr. Sauvage | SNEA(P) |
| Mr. de Soras | Bureau Veritas |
| Mr. Zumsteeg | CFEM |

The writing of the Guide was shared by several engineers, who contributed to the preparation of various chapters.

| | |
|---|---|
| Mr. Brozzetti | CTIM |
| Mr. Chabrolin | CTICM |
| Mr. Goyet | CTICM |
| Mr. Labeyrie | IFREMER |
| Mr. Putot | IFP |
| Mr. Recho | CTICM |
| Mr. Roche | SNEA(P) |
| Mr. Ryan | CTICM |

The Guide has been thoroughly reviewed and discussed, and its successive versions have accordingly been improved thanks to the comments and advice of Mr. Amiot (SNEA-P), Mr. Charleux (Bureau Veritas), Mr. de Leiris, General Engineer, and Professor Radenkovic (LMS, Ecole Polytechnique).

Mrs. Harnagea-Sirianu (CTICM) deserves special thanks for her assistance in the preparation of this Guide and for arranging its presentation.

The advice of CTICM, in the person of Mr. Ivor Ryan, on technical matters in the translation into English of the Guide is acknowledged.

# Contents

# PART I

## SCOPE AND
## FIELD OF APPLICATION
## OF THE GUIDE

# PART II

## ANALYSIS
## OF THE STATIC STRENGTH
## OF TUBULAR JOINTS

# PART III

# FATIGUE ANALYSIS
# OF TUBULAR JOINTS

# ANNEXES

# Notice

The reader's attention is drawn to the presentation of the Guide, which contains recommendations and commentaries.

In any given section, the commentaries appear after the recommendations.

To avoid confusion with the text of the recommendations, the text of the commentaries is indented and composed in italics.

# PART I

## SCOPE AND FIELD OF APPLICATION OF THE GUIDE

# CHAPTER 1

# Introduction

## 1.1 PURPOSE OF THE GUIDE

Construction codes for offshore structures require two types of analysis:

(a) Analysis concerning the action, or the combination of actions, whose amplitude has a low probability of occurrence during the service life of the structure. In other words, this analysis is conducted under the action of extreme forces. This generally leads to the analysis of the structure under a loading, or a combination of loads, occurring at a given moment (see Part II).

(b) Analysis concerning the action of repeated loads in time (wind, wave and, more rarely current). This is a fatigue analysis based on the knowledge of the distribution of these actions in time. In some cases, the calculation of the stresses generated by these actions may require the use of dynamic response calculations (see Part III).

Tubular joints by reason of their construction, necessarily display geometric discontinuities which are the sites of stress concentrations in zones located precisely in the neighborhood of the welds. Fatigue cracks are liable to be initiated and propagated in these zones.

*This guide is intended to draw the design engineer's attention to the construction and execution details of welded joints, and also to define the assumptions and methods concerning the calculation of the strength of tubular joints. Two types of joint analysis are discussed:*

*(a) The first relates to the action of extreme forces.*

*(b) The second relates to the action of dynamic forces, for which fatigue processes are liable to be generated (progressive damage by cracking).*

As a rule, it is impossible to identify the conditions or areas for which each type of analysis governs the design of the structure. The design engineer's attention is therefore drawn to the following remarks:

(a) Cumulative fatigue damage calculations for tubular joints should be preceded by a check on the punching shear strength under extreme loads (see Part II).

(b) As a part of the fatigue analysis of the joints of an offshore structure, it is essential to analyze systematically all the joints for which the fracture of a member terminating at this joint is liable to jeopardize the stability of the structure, either due to a loss of equilibrium, or because of a low reserve capacity to accomodate a redistribution of forces (this involves an assessment of the effect of "redundancy", following the fracture of the end of a member terminating in a joint).

**Example:**

Unless the structure is subjected to loadings which are only slightly variable, all the leg joints of a jacket must be analyzed for fatigue systematically, together with the joints forming part of the structure's stability system (bracing).

It is also recommended to check the fatigue of joints that are situated in:

(a) Places presenting very difficult accessibility to inspection, or where repairs raise special problems (all offshore repairs are problematic).

(b) The splash and tidal zone, and the pile penetration level.

## 1.2  DEFINITIONS AND NOTATIONS

### 1.2.1  TERMINOLOGY

Figure 1.1 gives the standard terms encountered in a tubular joint.

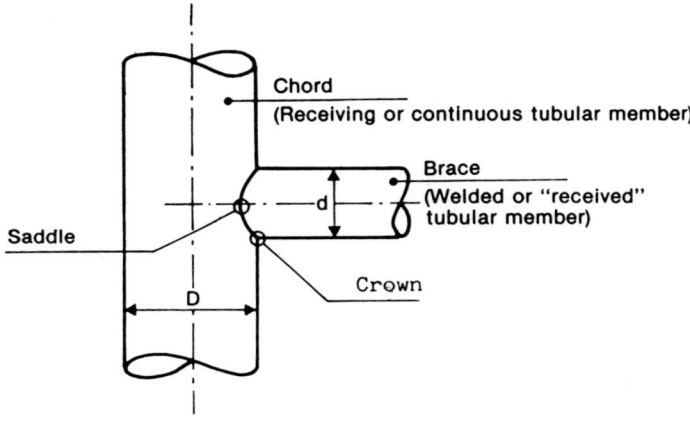

Fig. 1.1. Brace-chord connection.

*For the simpler joints of offshore structures, the saddle and crown points are important geometric points in whose zones stress concentrations occur under typical loading conditions. Hence these points are often the location of the "hot spots" discussed in Part III (Fatigue Analysis).*

## 1.2.2 NOTATIONS

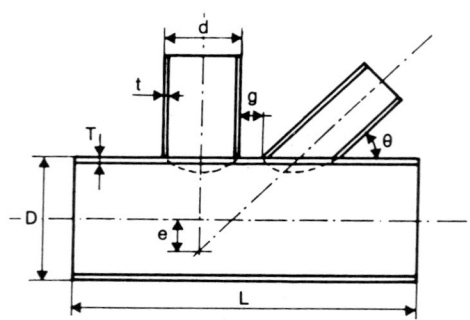

Fig. 1.2.

L = chord stub length,
D = chord outside diameter,
T = chord thickness,
d = brace outside diameter,
t = brace thickness,
g = theoretical gap,
e = eccentricity (positive in Fig. 1.2, negative otherwise),
$\Theta$ = acute angle defining the brace inclination,
$\alpha$ = 2L/D chord stub slenderness ratio,
$\beta$ = d/D brace to chord diameter ratio,
$\gamma$ = D/2T parameter defining the slenderness of the chord wall,
$\tau$ = t/T brace thickness to chord thickness ratio,
$\zeta$ = g/D relative gap.

In the case of two or more braces, they are identified by a subscript.

*The length L is used to calculate the stress concentra-tion factor given by parametric formulas for T and Y joints with the brace loaded axially.*

*The introduction of the parameter $\alpha$ = 2L/D in the para-metric formulas is primarly necessary for experimental considerations. The way in which L is selected in a real structure has never been fully clarified. However, this parameter exerts only a slight influence on the calcula-tion of the stress concentration factor (SCF) by parametric formulas. It was decided to use the stub length as the value of L, as defined in Section 2.1a.*

## 1.3  CLASSIFICATION OF TUBULAR JOINTS

### 1.3.1  SIMPLE GEOMETRY JOINTS

#### A.  T and Y joints

These joints feature a single brace perpendicular to the chord, or inclined to it (Fig. 1.3).

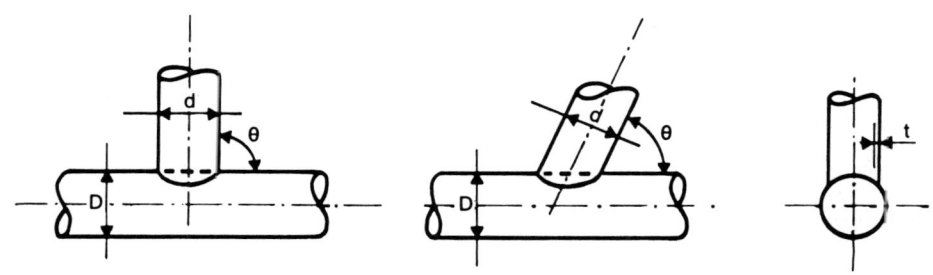

Fig. 1.3.

*A lower limit exists for the angle Θ (see Table 1.1).*

#### B.  X joints

X joints consist of two coaxial braces, on either side of the chord (Fig. 1.4).

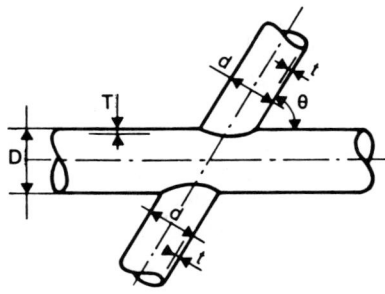

Fig. 1.4.

## C.  N, K and KT joints

These joints have two (or three) braces welded to the chord in the same plane (Fig. 1.5).

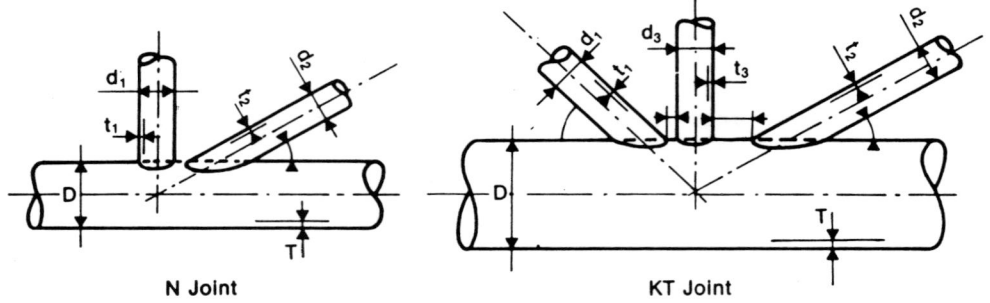

N Joint                          KT Joint

Fig. 1.5.

### Eccentricity

Three possibilities may occur as shown below (Fig. 1.6).

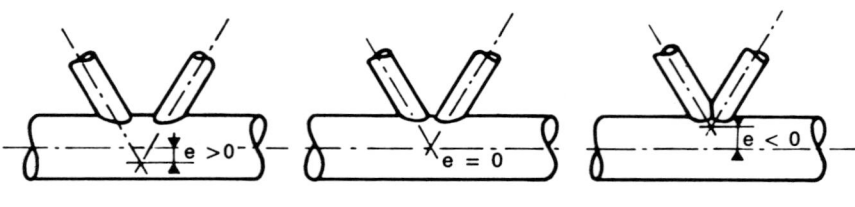

Fig. 1.6.

The conditions for which the secondary moment due to eccentricity should be taken into account in calculating the stresses are defined in Section 2.1c. The overlap joint is treated as a complex joint.

*As a rule, the centroidal axes of all connecting members should meet at the same point. For construction reasons, however (see Section 2.1), this convergence is not always feasible, or even preferable.*

### Gap and overlap

Overlap is stated to exist if two braces intersect before joining the chord, and a gap exists in the opposite case (Fig. 1.7).

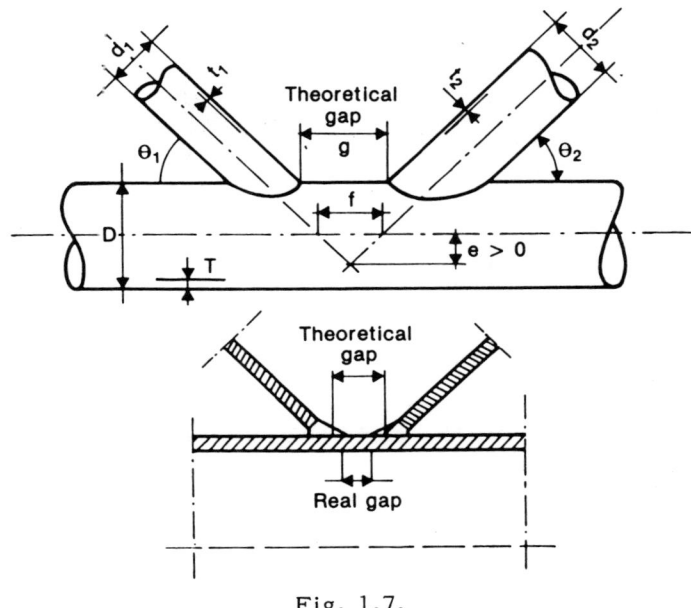

Fig. 1.7.

*Section 2.1d defines the conditions imposed on the real gap:*

*(a) K joints with overlap are classed with complex joints, due to the particular difficulty of evaluating the value of the stress concentration, as well as its position.*

*(b) Whatever the eccentricity sign, the centerline distance, measured on the chord axis, has the following value:*

$$f = e(\frac{1}{\tan \Theta_1} + \frac{1}{\tan \Theta_2})$$

**Typical values of geometric parameters**

Table 1.1 gives the values of the geometric parameters commonly encountered in unstiffened joints of offshore petroleum structures. It is recommended to adopt values lying in the typical range.

Table 1.1.

| PARAMETER | TYPICAL RANGE | MIN. VALUE | MAX. VALUE |
|-----------|---------------|------------|------------|
| $\beta = d/D$ | 0.4 to 0.8 | 0.2 | 1.0 |
| $\gamma = D/2T$ | 12 to 20 | 10 | 30 |
| $\tau = t/T$ | 0.3 to 0.7 | 0.2 | 1.0 |
| $\Theta$ degrees [1] | 40 to 90 | 30 | 90 |
| $\zeta = g/D$ | negative to + 0.15 | negative | 1.0 |

(1) For K, N and KT joints, the angle between two members should be:
. greater than 15° between any two braces,
. greater than 30° between a brace and the chord.

## 1.3.2  COMPLEX GEOMETRY JOINTS

With respect to joints with complex geometry, the design engineer's attention is mainly drawn to the lack of simple calculation methods which, in simple geometry joints, serve to determine the stress concentration factors or the static strength.

For complex geometry joints, the only methods available to analyze the local state of the stresses are:

(a) Numerical methods: finite elements analysis, for example.

(b) Experimental methods:
. Measurements on acrylic and epoxy models (by extensometric or photoelastic methods).
. Measurements on steel models: strains are normally obtained by extensometric methods.

Three main categories are distinguished among complex geometry joints:

(a) Overlap joints.

(b) Joints for which several tube connections exist on the same chord, located in one or more planes (other than those defined in Section 1.3.1).

(c) Stiffened joints.

*For simple geometry joints and for certain loading applications, the stress concentration factor can be calculated by simple parametric formulas. Moreover, the geometry and loading mode are such that the position of the hot spot can be identified with sufficient accuracy. This is not generally true of complex geometry joints.*

*A similar comment can be made about the evaluation of the static strength of joints of simple and of complex geometry.*

## A. Overlap joints

An overlap joint is a joint for which at least two braces intersect before joining the main chord.

No parametric formulas currently exist to calculate the stress concentration factor in K joints with overlap. The position of the hot spot varies considerably depending on which loaded brace is being considered.

As a rule, if the overlapping brace only is axially loaded, the hot spot occurs at the top of the overlap. If the through brace only is axially loaded, the hot spot is usually located somewhere in the chord/brace junction zone, but very rarely at the saddle or crown points.

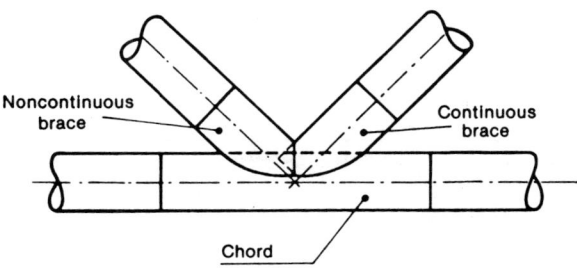

Fig. 1.8.

Overlap may occur if the braces and chord are in the same plane (Fig. 1.8), or if the braces and chord are in different planes (Fig. 1.9).

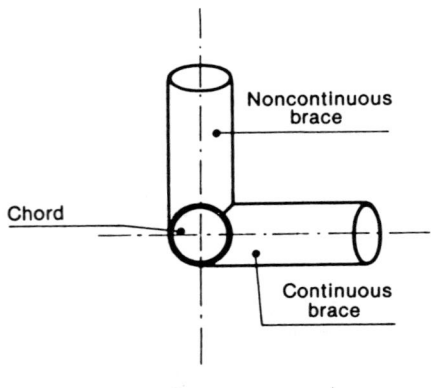

Fig. 1.9.

The conditions for obtaining the continuity or discontinuity of a brace are defined in Section 2.1c.

*Overlap ensures that the common welded cross-section of the overlapping braces withstand part of the shear force transmitted by the braces. Hence the cross-sectional area of the chord is not required to withstand the total shear force, and strains due to the shear force in the chord wall are thus limited.*

*The overlapping of one tube on another improves the bending rigidity of the chord wall. This local stiffening of the chord wall is due to the presence of a continuous brace/chord junction zone (Fig. 1.10).*

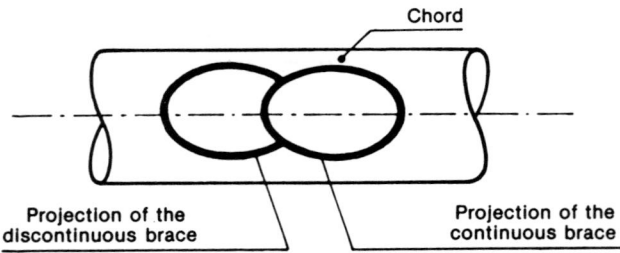

*Fig. 1.10*

*The foregoing remarks show the practical vaue of this type of joint (K with overlap) and explain the reasons why lower stress concentrations are observed than in K joints without overlap.*

## B. Joints with two or more tube connections

An example such a joint is shown in Fig. 1.11. Chapter 4, Part III, attempts to define the extent to which the stress concentration can be determined.

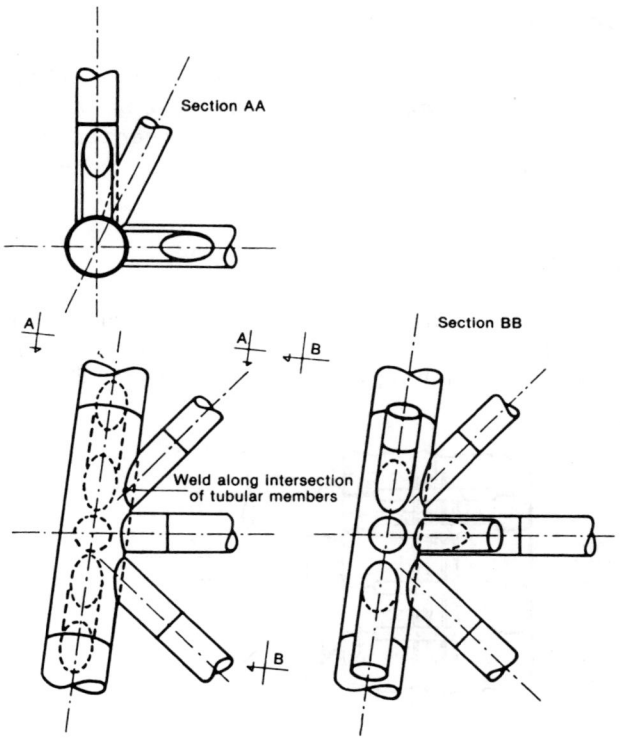

Fig. 1.11.

*By contrast with simple joints, the definition of a stress concentration factor applied to joints with two or more tube connections is ambiguous.*

*Case of simple loading (axial force or bending moments):*

In the absence of other clear and precise definitions, for each weld attaching a brace to the chord, one can, by convention, adopt a stress concentration factor relative to a simple loading for the purpose of fatigue analysis.

*Case of complex loadings (simultaneous presence of a normal force and bending moments):*

A clear definition of the stress concentration factor cannot be given for complex loadings. Part III, in Section 3.2.2, describes a safe method for calculating the variation in design stress.

This variation in design stress is evaluated from the superposition of stress concentrations relative to each simple force acting on the brace connected at the joint by the weld being analyzed.

## C. Joints with stiffeners

The purpose of stiffening the wall of a chord forming part of a tubular joint is to try to enhance the bending rigidity of the chord wall under the effect of the punching action of the brace (Fig. 1.12).

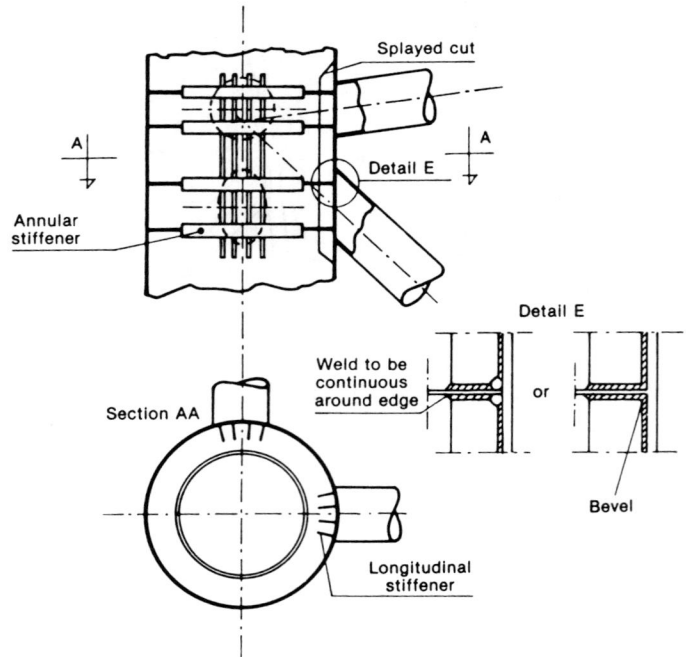

Fig. 1.12.

Longitudinal stiffeners serve mainly to reduce the value of the stress concentration due to a "simple" axial or bending force in the plane, and annular stiffeners serve to reduce the stress concentration value due to a simple bending force out-of-plane.

The continuity of annular stiffeners is preferably maintained, making the longitudinal stiffeners discontinuous. Longitudinal stiffeners are welded to the annular stiffeners to restore continuity. The annular stiffener may consist of several lengths as shown in Section AA of Fig. 1.12.

The use of stiffeners, due to the welds they contain, inevitably creates new stress concentration zones. If care is not exercised, the real improvement they offer in stress concentrations at the tube intersections may be offset by the new risks that they thus incur.

Apart from specific cases requiring special investigation, construction details involving external stiffening are not recommended for offshore petroleum structures.

*In non-offshore tubular structures, the use of tubes of small diameter prevents the use of internal stiffeners.*

*Stiffening the continuous chord of the joint substantially increases punching strength. In the external stiffening method, the stiffeners are also fixed to the braces. The forces transmitted to the chord by the braces pass through the external stiffeners, thus generating stress concentrations located at the stiffener/brace junctions.*

*In this external stiffening method, while the punching strength is substantially improved, no improvement in fatigue strength is achieved. Fatigue strength may even tend to drop sharply, due to the premature appearance of fatigue cracks in the critical stress concentration zones (Fig. 1.13).*

*Experimental tests have revealed the possibility of fatigue cracks occurring at the weld attaching these stiffeners, so that attention must be paid not only to the execution of the welds, but also to calculating the local stresses in the neighborhood of these welds connecting them to the main elements, chords and braces.*

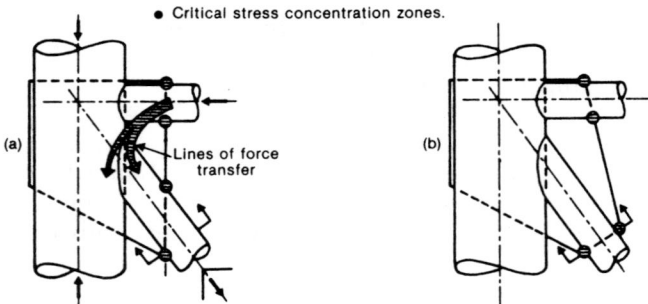

Fig. 1.13

As a rule, on fixed structures, it is impossible to inspect and repair internal stiffeners in a structure in service.

# CHAPTER 2

# General Guidelines
# on Joint Details

The following two Sections, Sections 2.1 and 2.2, review some of the conditions associated with the geometry and preparation of joints and welds.

Section 2.2 in particular sets forth precise requirements concerning the weld profile in the neighborhood of the toes. The fatigue service life is partly conditioned by the real local profile radius at the weld toes at the junction of the weld and the wall of the chord and brace. Proper execution of this detail serves to increase the time before crack initiation.

Section 7.2 of Part III introduces the concept of "inspected" weld profile, in the sense of inspection after execution, as well as the application of the precise requirements in Section 2.2.

*The fatigue strength of an offshore structure joint is very strongly influenced by the preparation, fabrication and finished condition of the tubes to be assembled, and also the quality control of the welds.*

*Part II merely defines a number of conditions concerning:*

*(a) Tube cutting and edge bevelling.*

*(b) Tube connection and tack welding.*

*(c) Weld geometry and dimensions.*

*Other highly important conditions are liable to play a significant role in improving weld execution conditions:*

*(a) Choice of steel grades and qualities, combined with the choice of welding procedures and parameters.*

*(b) Welder qualification.*

*(c) Quality control and inspection.*

*Since the latter two matters are covered by specific code requirements, they are not dealt with in this guide.*

## 2.1  GEOMETRIC CONDITIONS FOR THE PREPARATION CUTTING AND BUTT WELDING OF TUBES IN FABRICATING A JOINT

*The distances to the stub/tube butt welding planes given as recommendations are designed to keep this welded zone away from the stress concentration zones. For large joints, involving the welding of very thick plates (see Section 3.7.4) the need for overall stress relief heat treatment of the joint may arise. If so, practical stub length requirements given as recommendations may lead to joint sizes exceeding the clearances of heat treatment furnaces. This might mean reducing the stub lengths, but in so doing it is important to ensure that the butt welding plane lies outside a high stress concentration zone.*

(a) If the chord thickness must be increased locally at a joint, the distance between the chord/stub butt joint plane and the nearest chord/brace intersection point must be at least equal to the greater of the following two values:

   (1) 300 mm.

   (2) One quarter of the chord diameter.

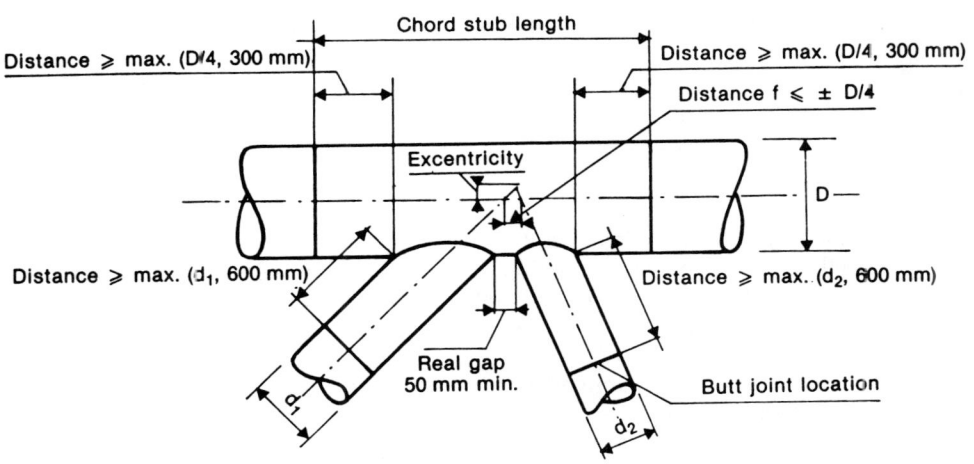

Fig. 2.1.

The same rule applies if a different steel grade is used for the chord-stub than for the remainder of the chord.

> *To improve the static and fatigue strength of a joint, the chord-stub thickness may be increased (see Parts II and III on the effect of the parameter $\gamma = D/2T$). If this leads to high thicknesses giving rise to fabrication problems, a higher yield strength steel is used for the stub, which however achieves an improvement in static strength only.*

(b) Where increased wall thickness or a special steel is used for braces in the chord/brace joint area, the distance between the brace-stub butt joint line and the nearest brace-stub intersection point on the chord must be at least equal to the greater of:

    (a) 600 mm.

    (b) One quarter of the brace diameter.

(c) In a joint of two or more braces with a chord, the distance f between the intersections of the centerlines of these braces with the chord centerline should not exceed one quarter of the chord diameter. Where this requirement cannot be achieved for construction reasons, the secondary moment resulting from eccentricity should be considered in the structural analysis (Fig. 2.1).

(d) The real gap should not be less than 50 mm.

> *The definition of the real and theoretical gaps is given in Section 1.3.1b.*

(e)   For K joints with overlap:
- The sizing of the overlap (i.e. of the weld between two braces) should be such as to withstand at least 50% of the component of the axial force N perpendicular to the chord centerline (Fig. 2.2).
- The thicker or more highly stressed brace (through brace) is welded continuously to the chord: the thickness of this brace should never exceed the chord thickness.

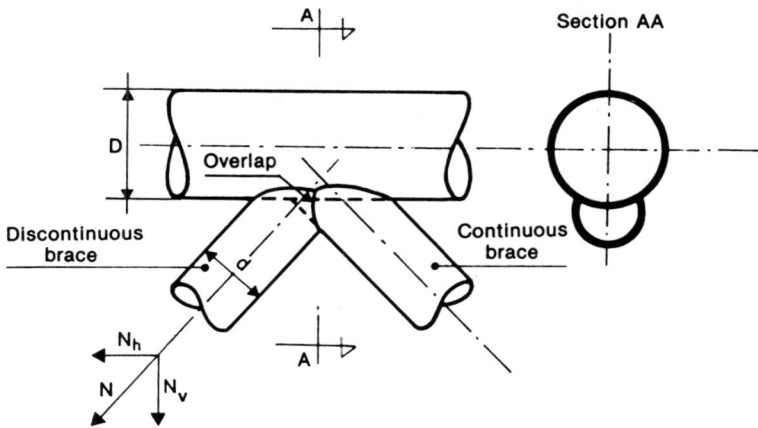

Fig. 2.2.

(f)  The butt joint distances includes bevels in the case of the jointing of tubes of different thicknesses.  Tubes generally have the same outside diameter in offshore petroleum structures. Fig. 2.3 gives the butt welding conditions for tubes of different thicknesses and the same outside diameter.

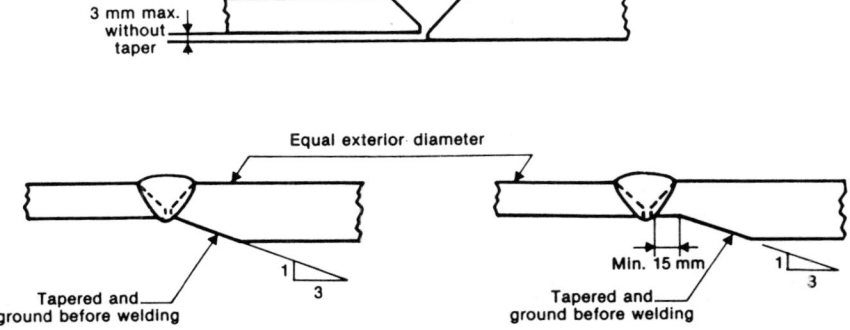

Fig. 2.3. Details for butt joints welded from one side only - No sealing run possible.

## 2.2  CONDITIONS GOVERNING THE PREPARATION OF TUBE EDGES AND WELD PROFILE

The requirements given in Fig. 2.4 correspond to the case in which welding is carried out from one side only.

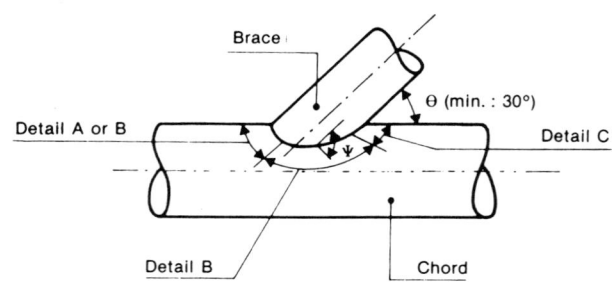

Weld preparation and weld profiles

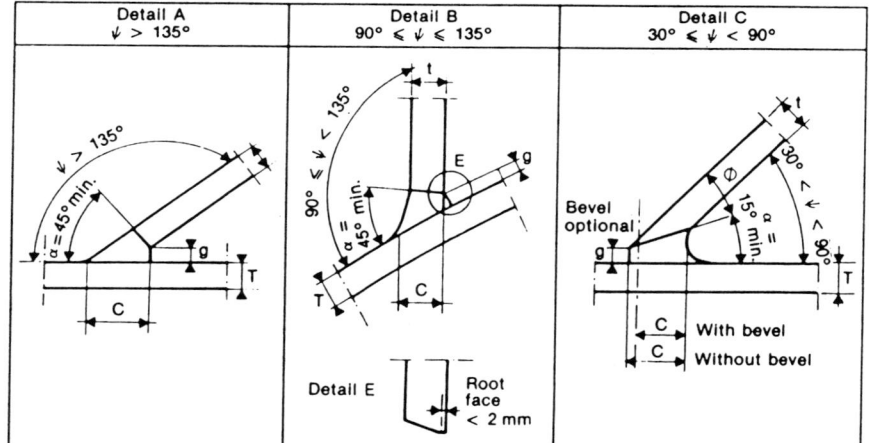

Table 2.1

| Detail | Dihedral angle | Min. C |
|---|---|---|
| A | ψ > 135° | min. $\begin{cases} 1.75\,t \\ t/\text{Sin}\,ψ \end{cases}$ |
| B | 90° ≤ ψ ≤ 135° | 1.25 t |
| C | 50° ≤ ψ < 90° | 1.25 t |
| | 35° ≤ ψ < 50° | 1.50 t |
| | 30° ≤ ψ < 35° | 1.75 t |

Table 2.2

| Opening angle α | Root opening g (mm) |
|---|---|
| α < 45° | 3.0 – 6.5 |
| 45° ≤ α ≤ 90° | 1.5 – 5.0 |
| α > 90° | 0.0 – 5.0 |

Fig. 2.4.

At the brace to chord joints, the root opening distance g defined in Table 2.2 of Fig. 2.4 should be respected.    This distance is provided by wedging and tack welding.

The distance from the root to the weld toe is given in Table 2.1 of Fig. 2.4 as a function of angle $\Psi$.

The tolerance of the bevel angle ($\varphi$) ($\varphi = \Psi - \alpha$) is $\pm 5°$. If the cut is reclosed due to differential shrinkage resulting from the welding operation, the cut can be re-opened by arc gouging, to satisfy the groove conditions shown in Fig. 2.5.

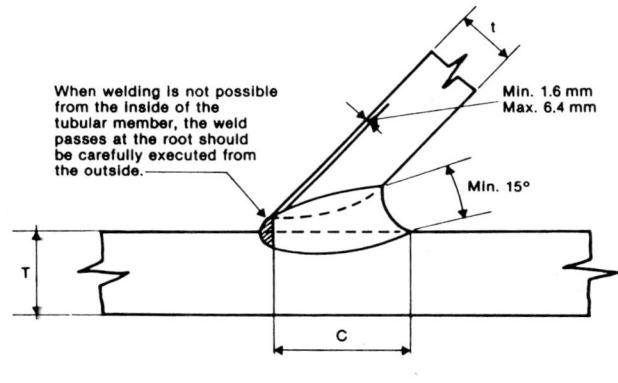

Fig. 2.5.

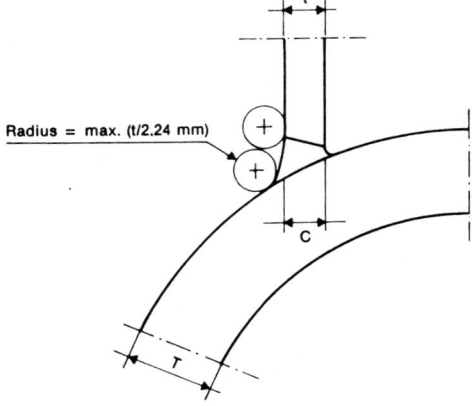

Fig. 2.6.

The acceptability conditions for an "inspected" weld profile (Fig. 2.6) are given in Part III, Chapter 7.

*The intersection of two (or more) tubes forms a tubular joint at which stress concentrations occur. Part III, Section 2.2.2 draws the design engineer's attention to the fact that these stress concentrations are strongly influenced by the local weld geometry emphasizing the importance of the care to be applied in the execution of the weld to tube-wall transitions at the intersection.*

*If the members are relatively thick, spot welds for tack welding are executed with a smaller diameter electrode than those used in the main welding passes, to ensure their penetration to the root of the cut. These spot welds should be sufficently large (e.g. 20 mm) to withstand the deformation of the members during welding.*

CHAPTER **3**

# Steel Grade Choice
# for Welded Steel Structures

## 3.1 INTRODUCTION

The steel grades required for the construction of a welded component of an offshore structure depend largely on the importance of this component in the safety of the structure. Methods for checking the strength of structures, which refer to the metal's yield strength, implicitly assume that the maximum strength of the components is synonymous with ductile failure. Therefore it is also necessary to make sure that brittle fracture cannot occur during the lifetime of the structure. To do this, the designer must check that:

(a) The fracture toughness of the steel selected is sufficent, given the conditions in which the structure will be placed; the methods for selecting steel quality (Section 3.5) are aimed to achieve this.

(b) On completion of the different fabrications operations, the steel should still display the impact strength specified by the regulations and codes; this is the purpose of the requirements concerning welding and forming methods, and any heat treatments applied (Section 3.6 and Section 3.7).

The conditions to be taken into account in selecting steel quality are all those which may exert a detrimental effect on the strength of the components with respect to brittle fracture. These are associated with:

(a) The service conditions of the structure: temperature and the extent to which fatigue type loading occurs.

(b) The design of the structure: thickness of parts and stress magnitudes.

(c) Execution procedures: the inevitable presence of certain defects in welded joints.

It is at low service temperatures that the risk of brittle fracture is the greatest. The unfavorable effect of the forces increases with their magnitude and loading rate.

Experience, and more recently, certain fracture mechanics methods, have helped to establish correlations between the values of these different parameters and the occurrence of brittle fracture. These correlations provide the basis of methods for selecting steel qualities discussed in Section 3.5.

*Extremely simple steel selection criteria for the design of structures of the industrial type have been employed for a very long time. The tensile strength served as a design basis, while the minimum elongation at rupture guaranteed the ductility of the material. This led to the development of adaptations that were made inevitable by the simplicity of the joint design methods.*

*The introduction of welding as an assembly procedure completely altered this situation. The main property of a steel became its in-service behavior, primarly with respect to the risk of brittle fracture - on which attention was focused following a number of spectacular accidents - and then, more recently, with respect to fatigue failure.*

*By hindsigth, one can now assert that the criteria adopted to forestall these risks were basically highly empirical. In these conditions the statutory texts could not always escape the accumulation of requirements, with variable justification.*

*Fortunately the vast body of research conducted has culminated in a situation where, at least for the most commonly used steels, all the data are available to build quality structures. However, the requirements of the various documents and specifications sometimes reflect the diversity and the uneven value of the work undertaken, and are mired in a certain complexity, redundancy, and indeed, sometimes, certain disputable aspects which perpetuate requirements that are no longer in tune with the growth of knowledge.*

*The purpose of Section 3 is to compile the concepts which form a reliable body of doctrine on structural steels, their properties and behavior, and on the weldabililty conditions, and above all, to refer to the basic documents which, in the future, should either progressively supplant the texts that are now obsolete, or should serve as references.*

*For the case of welded steel structures subject to very severe environmental conditions (Arctic zone) or situated in very cold regions, more stringent criteria must be considered than those pointed out in this text.*

*The choice of steel grades could logically give rise in a given structure to the existence of different qualities, in accordance with the welding processes and procedures, with the construction details, and, depending on the severity of the forces encountered in service in a given part of the structure. For a given structure, however, it is obviously in the builder's interest to select a series of steel qualities in a range offering a reasonable spacing (sufficiently wide difference between two consecutive qualitites).*

## 3.2  PROPERTIES OF STEELS.  A SELECTIVE REVIEW

This Section reviews a number of concepts concerning the properties of steels (the base material) and the welding of these steels.

### 3.2.1  THE BASE MATERIAL: STEEL

#### A.  Structural steel grades

The grade of a steel - in the sense of the standards defining the grades and qualities of structural steels - refers to the tensile properties of the metal: yield strength, tensile strength and elongation.

In French standards, the grade is expressed by an index equal to the minimum guaranteed yield stress $R_{eG}$, usually expressed in megapascals (MPa), equivalent to Newtons per square millimetre (N/mm$^2$).

*Reference: [3.1].*

#### B.  Steel quality: fracture toughness

The quality of a steel - in the sense of the standards defining steel grades and qualities - essentially concerns its fracture toughness, in other words its ability to resist brittle fracture.

The fracture toughness of the steel is not an immutable property. It also depends on the fabrication methods implemented in the steel mill, and may be altered by the mechanical (bending, shaping, etc.) and thermal (welding) operations in fabrication.

*Reference: [3.2].*

#### C.  Quality criteria

The fracture toughness of a steel is determined by an impact bend test on a notched specimen.  The test result is expressed, with an indication of test temperature, by the fracture energy (in joules) or the energy absorbed per unit area (in joules/cm$^2$).  In the latter case, the result is called the resilience of the metal, whereas the former in known as the fracture toughness.

This result depends on the types of specimen and notch employed. In this guide, reference is made exclusively to the test on a V-notch specimen (Charpy V test specimen).

*The impact bend test on a simply supported V-notch specimen is defined by Standard NF A 02-161.*

*The symbol for fracture toughness is KV, and that of resilience is KCV.*

*The term "resilience" is sometimes used abusively, to describe the "fracture toughness" of the material.*

*AFNOR Standards maintain the reference to the energy absorbed per unit area (i.e. resilience) although in practice its use is rare.*

## 3.2.2  WELDING

### A.  Weldability

The weldability of a steel is a complex property which depends as much on the material's intrinsic properties as on the shape of the parts and the way in which welding operations are conducted. This is why they can only be checked after complete definition of the structure and the welding procedures, and only on samples prepared with the steel employed which reproduce the construction details selected (dimensional characteristics, restraint conditions of the parts, and welding procedures specified in fabrication).

### B.  Weldability criteria, cold cracking

Despite the complex nature of weldability, failing accurate measurement, an attempt is made to evaluate it as closely as possible from criteria that are easy to use. Of all the criteria proposed, two retain a definite practical interest, although their validity is far from absolute:

(a) Carbon equivalent.

(b) Weld hardness.

Close attention should be paid to cold cracking, which is known [3.3] to require the presence of three essential factors in order to occur:

(a) The existence of stresses applied to the welded joint.

(b) The presence of hydrogen introduced into the heat affected zone during welding.

(c) The existence in this zone of metallurgical structures, liable to hydrogen embrittlement.

This guide recognizes as a basic principle that the quality of a steel for welding cannot be selected without the knowledge of the welding procedure. On the subject, ref. to Section 3.7.1.

### Carbon equivalent

*Many experimental formulas are available to determine the maximum hardness of the heat affected zone (HAZ) or the base metal cold cracking sensitivity. The best known formula is the one recommended by the International Institute of Welding (IIW), which was established for steels with relatively high carbon content, i.e. in the neighborhood of 0.40%).*

*The carbon equivalent calculated by the equation recommended by the IIW*

$$Ce = C + \frac{Mn}{6} + \frac{Cr + Mo + V}{5} + \frac{Ni + Cu}{15}$$

*C, Mn, Cr, ... the carbon, manganese, chromium ... contents of the steel in %*

*is a first approach to dealing with the risk of finding constituents sensitive to embrittlement in the HAZ under the effect of welding. Obviously the presence of these constituents depends on the steel composition, but many other factors are involved, whose effects cannot be identified and controlled by the conditions imposed on composition alone.*

*It is clear that, all other things remaining equal, variations in composition can be interpreted. Yet the results anticipated from this expedient are only meaningful for the type of steel for which they have been established. This is why it is always surprising to find a single value of Ce imposed in some cases on a whole range of steels whose only common feature is their mechanical properties, but which differ in composition and structure. Note that a low value of Ce does not necessarily mean that one is on the safe side.*

### Weld hardness

*In the past, weld hardness was used to characterize the hardness-cooling parameter curves in a simple way. Here also, the imposition of a single hardness value for steels with different compositions and structures is no longer justified, and one can only hope that this practice will be abandoned. It is inconsistent to expect to obtain low hardnesses together with high resilence.*

## C. Lamellar tearing

Lamellar tearing is another serious defect liable to affect welded joints. This is the separation by cracking that occurs when the weld shrinkage forces are exerted across the thickness of the part, on a zone of the base metal that is weakened by the presence of nonmetallic inclusions.

The sensitivity of a steel to this defect is evaluated by measuring the reduction in area during a tensile test on a through thickness specimen (see Section 3.7.5).

*Reference:* [3.19].

## D. Steel strain hardening and ageing

The ageing of a steel generally occurs after cold work hardening (shaping, punching, rolling), which results in the deterioration of the properties of the steel, materialized by an increase in the strength properties, a decrease in ductility and an upward shift in the transition temperature. Depending on the grades and the initial properties, this means that the steel may become brittle at the service temperature.

Ageing is also accelerated by an increase in temperature, as in steel plates that are work hardened during mechanical operations, and then raised to elevated temperatures by welding operations.

Section 3.6 reviews the requirements concerning plate metal shaping.

*Reference:* [3.5].

## 3.3  PRODUCT CONTROL AT THE STEEL SUPPLIER

If the offshore structure concerned has to be "classed" or "certified" by a classification Society, or built under the supervision of an inspection agency designated by the client or by the country in whose territorial waters the structure will be placed, controls at the steel production plant must be planned at the time of ordering, in accordance with the relevant regulations and requirements of the classification Society or of the inspection agency responsible for supervising construction.

In this case, the order, in the "acceptance" Chapter, (in this respect see Standard NF A 03-115 for the conditions of delivery of steels, and NF A 49-000 for the conditions of delivery of steel pipes and tubular products) must specify:

- The name of the classification Society or the inspection agency responsible for works inspection.
- Quantity (number of pieces or weight).
- Shape of the product.
- Nominal dimensions.
- Dimensional tolerances.
- Steel grades and qualities.
- Agreed utilization guarantees.
- Technical requirements.
- Heat treatment conditions.
- Surface treatment and finishing conditions.
- Types and conditions of inspection to be performed on each batch, and the types of control document to be furnished.
- Product marking conditions, finishing, packaging, loading and shipment conditions and the destination.

*The foregoing details can be defined by reference to:*

*(a) Standards.*

*(b) Regulations [ 3.10, 3.7, 3.8 ] or detailed specifications of oil companies [3.6, 3.15].*

*After inspection by the classification Society or the inspection agency, the products are stamped and an inspection certificate is prepared in conditions similar to those listed in the relevant Sections of the Rules for Materials of Bureau Veritas [3.10].*

*The control documents enabling the producers of steel products to communicate to their clients the results of the controls performed in the works are defined in the following standards:*

*NF A 03-116 (Steel products),*

*NF A 49-001 (Steel pipes and tubular products).*

## 3.4  CLASSIFICATION BY CATEGORY OF WELDED STRUCTURAL ELEMENTS

Welded steel structural elements are classified in three strength categories:

(1) Special category.

(2) First category.

(3) Second category.

These categories are distinguished by the importance of the consequences in terms of the safeguarding of human lives and economic losses that could result from the collapse of the structure or of one of its components.

### Special category elements

Elements in this category are those highly stressed first category structural elements whose complicated shapes and structural details incur uncertainities in terms of design, construction or inspection.

### First category elements

The first category includes structural elements whose failure incurs the risk of collapse of the structure, or whose failure could cause the disabling of the platform.

### Second category elements

The second category includes structural elements that do not fall into the foregoing two categories.

In preparing working drawings, it is recommended to note the relevant category for each structural element  and especially for the joints. Similarly, it is important for the drawings to mention, with a notation as complete as possible, all details concerning the choice of the steel quality and the fabrication conditions including:

(a) Designation of the steel grade.

(b) Charpy V test conditions (direction of sampling of test specimens, impact test temperatures, value of fracture energy).

and if applicable:

(a) Guaranteed reduction in area in the short transverse direction.

(b) Guaranteed fracture toughness after ageing test.

(c) Preheat conditions.

(d) Stress relief heat treatment.

*Tubular joints are classified as described below.*

*Special category:*

   *(a) Jacket leg joints (main chord stub and brace stub).*

   *(b) Joints that are highly stressed by concentrated loads (such as lifting, or at the time of launching).*

   *(c) Joints located in the splash zone.*

   *(d) Parts connecting the deck to highly stressed legs.*

*First category:*

   *(a) Joints of uprights and diagonals not forming part of the jacket legs.*

   *(b) Joints of support and guide structures for the risers and conductor pipes (grids).*

   *(c) Braces of the jacket and deck.*

*The fact that these joints are classed in the first category in no way waives the need to check their fatigue behavior.*

*Second category:*

*Note that joints of "jacket" structures are never classed in the second category.*

## 3.5  METHOD FOR SELECTING STEEL GRADES

The choice of the fracture toughness test temperature of a steel for the construction of an offshore structural element depends on:

. The category of the member (Section 3.4).
. The design temperature (Section 3.5.1).
. The thickness of the material employed (plates, welded tubes and seamless tubes).
. The steel grade selected (yield strength).
. Any heat treatment after fabrication.

*Apart from specific structural elements (racks in jackup platforms, etc), in offshore petroleum structures, the steel grade is generally limited to the choice of steels with yield strengths less than 420 MPa (unalloyed manganese carbon steels, or weldable low alloy steels) where the thickness of the products employed (plate or tubes) are less than 80 mm.*

*Standard NF A 36-212 currently at the public enquiry stage and intended to supplement Standards NF A 35-501 and NF A 36-210 related to structural steels for routine use, concerns the definition of the grades and qualities of plate for the fabrication of offshore components designed to withstand especially severe operating conditions (North Sea for example).*

*This standard defines a number of characteristics:*

*(a) The guaranteed mechanical properties (tensile strength, bend test, fracture toughness, reduction in area across the thickness, internal soundness).*

*(b) Guarantees of ladle chemical composition and carbon equivalent values of the product for the qualities of grade PF36.*

## 3.5.1  DEFINITION OF DESIGN TEMPERATURE ($T_D$)

For offshore petroleum structures, a distinction must be drawn between the exposed part (part of the structure in the air, or in the splash zone) and the submerged part (part of the structure permanently underwater).

For the submerged part, the design temperature is taken by convention as 0°C. In temperate and warm zones, however, it is accepted that the design temperature is the water temperature on the coldest day of the year in the area where the structure will be used.

*In the absence of precise statistics, the table below gives the recommended design temperatures for three different offshore climatic zones [3.6].*

| Climatic zone (2) | Design temperatures $T_D$ exposed part | Examples of sites |
|---|---|---|
| 1<br><br>Cold seas | $-15° \leqq T_D < 0°C$<br>take $T_D = -15°C$ | North Sea (1)<br>Baltic Sea<br>Irish Sea |
| 2<br><br>Temperate seas | $0°C \leqq T_D < 15°C$<br>take $T_D = 0°C$ | English Channel<br>Bay of Biscay<br>Western Mediter-ranean<br>Eastern Mediter-ranean |
| 3<br><br>Warm seas | $T_D \geq 15°C$<br>take $T_D = 15°C$ | Gulf of Guinea<br>Persian Gulf<br>Indonesia<br>Red Sea |

*(1) In the North Sea, the Department of Energy recommends using the service temperatures of -10°C for the atmospheric zone and +4°C for the submerged zone.*

*(2) Climates related to seas close to the Arctic Ocean (North of the 65th parallel) are not included in this table.*

## 3.5.2 PRINCIPLE OF THE SELECTION METHOD

The principle of the selection method consists in determining the temperature at which the energy absorbed KV during an impact test on a V notch specimen will have the minimum conventional values set by the table below (in joules).

I  Steels for which $R_{eG} < 300$ MPa

|  | Special category | 1st category | 2nd category |
|---|---|---|---|
| Mean minimum value for 3 tests (J) | 27 (L)<br>20 (T) | 27 (L) | 27 (L) |
| Individual minimum value | 18 (L)<br>14 (T) | 18 (L) | 18 (L) |

II  Steels for which $300$ MPa $\leqq R_{eG} \leqq 420$ MPa

|  | Special category | 1st category | 2nd category |
|---|---|---|---|
| Mean minimum value for 3 tests (J) | 34 (L)<br>24 (T) | 34 (L) | 34 (L) |
| Individual minimum value (J) | 22 (L) | 22 (L) | 22 (L) |

(L) Test specimen taken parallel to the rolling direction.

(T) Test specimen taken perpendicular to the rolling direction.

Note:

(1) The values given in these tables concern finished products after fabrication (Section 3.7.3).

(2) For joints, it is recommended to record the values of the fracture energy on test specimens taken in the transverse (through thickness) direction.

*The minimum Charpy V fracture energies given in the recommendations are those appearing in the Rules for Materials of Bureau Veritas 1980 Edition, Chapter 2, Section 2.2 (update N° 1, January 1983). The steels have been divided into two groups according to the minimum guaranteed value required. Some regulations and recommendations [3.8] propose retaining the same minimum energy irrespective of steel grade, but require a lower impact test temperature for grades with higher characteristics ($R_{eG} > 320$ MPa).*

### 3.5.3  TEST TEMPERATURE

The test temperature on a V notch specimen is determined from the general diagrams given below. These diagrams, which help to select steel grades for the construction of offshore petroleum structures, give the temperature at which the Charpy V fracture energy values must be guaranteed to ensure, in principle, a sufficiently low probability of fracture of a structure in service. For a given minimum fracture energy (see Section 3.5.2), this temperature depends on the design temperature and the thickness of the steel product employed (tube, plate, etc.) (Diagrams I, II and III).

**Special Category**
**As welded condition without post-weld heat treatment for steels with $Re_G \leqslant 420$ N/mm²**

For $T_o < -20°C$ the steel grade should be decided in consultation with the classification society

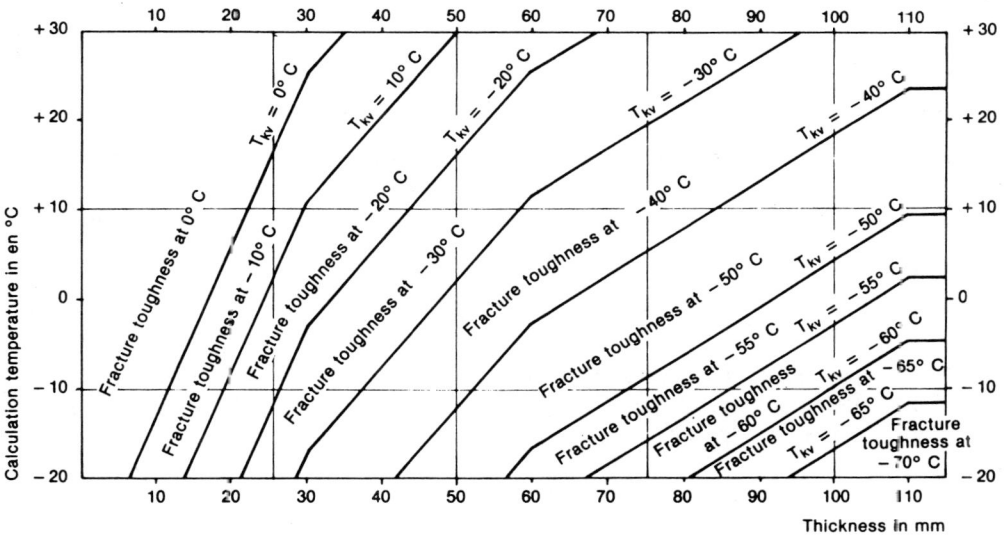

Diagram I (extract from reference [3.10])

**1st Category**
**As welded condition without post-welded heat treatment**

**Special Category**
**With post weld heat treatment for steels with $Re_G \leqslant 420/N/mm^2$**

For $T_o < -20°C$ the steel grade should be decided in consultation with the classification society

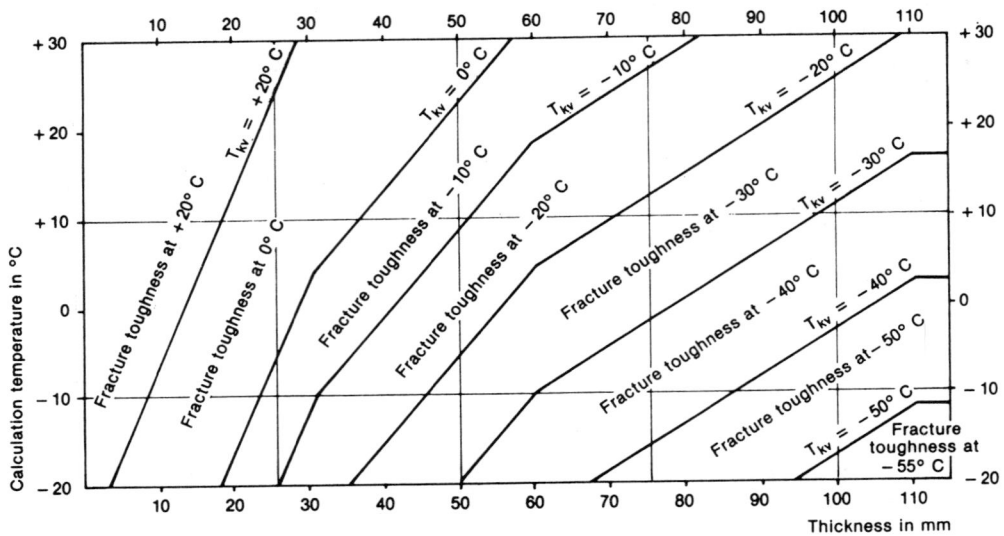

Diagram II (extract from reference [3.10])

**2nd Category**
**For steels with $Re_G \leqslant 420 \ N/mm^2$**

For $T_o < -20°C$ the steel grade should be decided in consultation with the classification society

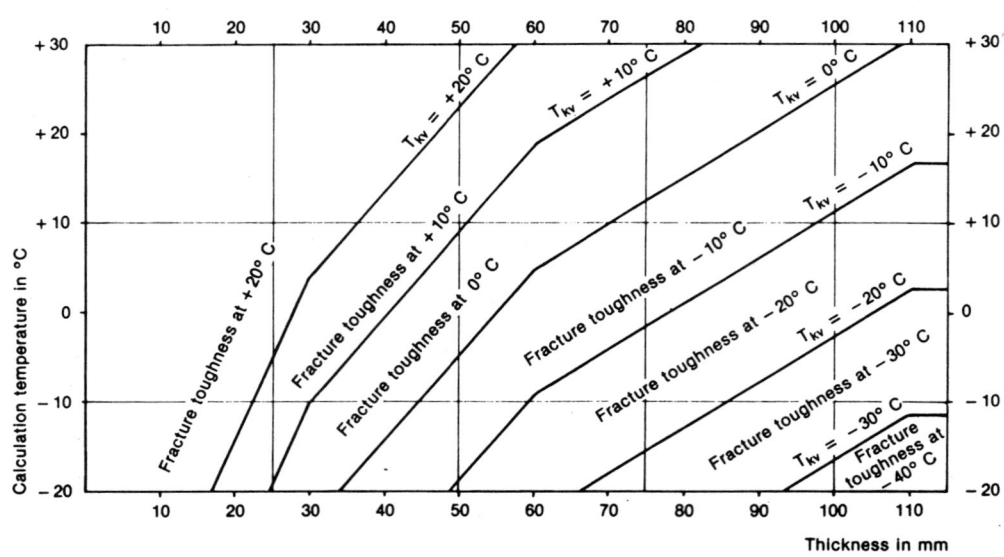

Diagram III (extract from reference [3.10])

*The justifications and assumptions underlying the determination of the diagrams given in the recommendations are given in Ref. [3.9] and in document NF A 36-010. These diagrams have been adapted to the choice of steel qualities of the structures of offshore platforms in update N° 1, August 1982 of the Regulations of Bureau Veritas for the construction and classification of offshore platforms, Chapter 5.*

*This general approach, which serves to determine the choice of steel qualities depends on many parameters, and it is important to keep in mind that, for a welded structure, only a limited number of steel grades should be considered, which possess clearly defined properties and are justifiably appropriate.*

*The comparison between diagrams I and II identifies the difference in requirements relative to the special category between a steel in the as welded state and a welded steel with a stress relief heat treatment. For conventional low alloy carbon steels, the stress relieving heat treatment has the effect of substantially reducing the residual stresses. This is why the requirements concerning the impact test temperature are less severe in this case.*

*In this respect, the fracture toughness and cold cracking strength can be improved by reducing the carbon content, but this is generally detrimental to the steel's tensile strength. This drawback is overcome in different ways:*

*(a) By adjusting the chemical composition by the addition of microalloy elements.*

*(b) By quenching and tempering treatment.*

*(c) By single controlled rolling.*

*(d) By different sorts of controlled thermomechanical rolling.*

*It should be noted that during welding, or during stress relief heat treatment, the microstructure (and/or tensile strength) of the base metal may be destroyed, so that the fracture toughness of these types of steel decreases significantly. It is important to make sure that the conditions pertaining to the fabrication of steels obtained by special techniques are complied with by the steel makers.*

*For the time being, it is not always sufficient to impose minimum guaranteed Charpy V impact values only. Certain foreign regulations governing the construction of offshore petroleum structures [3.7, 3.8] require a minimum critical crack opening displacement (COD) at the root, especially for thick steel products (base metal and filler metals for welding). These regulations consider that the value of the COD provides a better appreciation of the maximum allowable defect sizes.*

*A document, published by commission X "Residual stresses and stress relieving" of the International Institute of Welding, called "Reservations with respect to the application of elastic-plastic fracture mechanics to welded structures" IIS/IIW-707-82, Volume 20, N° 7/8 of "Welding in the World", expresses serious reservations about the sound basis of this approach. In this respect, note also the remark made in Ref. [3.9].*

*"The agreement between real sizes and calculated sizes of defects is not excellent, and this concept of the COD provides a sufficient condition of non-fracture for the time being only. Although it errs on the safe side, the discrepancy between the method and test results reach values which are too great for this type of analysis to be universally acceptable."*

## 3.6  PLATE FORMING

### 3.6.1  COLD FORMING BY ROLLING

The ageing of steel occurs after cold work hardening. It results in the progressive deterioration of its toughness. Ageing can be accelerated if work hardening is followed by moderate heating. This is to be feared, especially for steels that are cold formed and then weld fabricated.

The cold forming conditions of plates used to build offshore structures must therefore be subjected to a qualification procedure. The important aspects of the qualification procedure for steels for which $R_{eG} < 420$ MPa are reviewed below:

Calculate the strain ratio by the equation:

$$A\% = \frac{t}{d_i + t} \times 100$$

where

t = plate thickness,

$d_i$ = inside diameter.

If A% $\leq$ 5%: no qualification procedure is required.

If A% > 5%:
- Form a test coupon in the planned fabrication conditions.
- From the test coupon, machine specimens with a V notch, subject them to simulated ageing by heating at 250°C for 1 h.
- If the fracture toughness remains at least equal to the minimum value imposed by the Regulation or by the Building Code, no heat treatment is necessary.
- If not, proceed with stress relieving heat treatment and, on the sample, check that the fracture toughness of the metal after forming and treatment is at least equal to the minimum fracture toughness imposed by the Regulation or by the Building Code.

Consult Refs. [3.4 and 3.5 ].


*The equation given in the recommendations for the strain ratio is only valid for forming by rolling, so that for other fabrication techniques, it is necessary to use a modified equation adapted to the specific case concerned.*

## 3.6.2  WARM FORMING

Forming is said to be "warm" if conducted at a temperature between 150°C and 550°C. The warm forming conditions must pass a preliminary qualification procedure.

*The maximum temperature not to be exceeded is $AC_1$. Nevertheless heating temperatures up to 650°C and even 700°C can be envisaged as from now. To avoid the deterioration of certain characteristics, as well as the risks of embrittlement in forming, it appears advisable to keep the forming temperature above 450 to 500°C.*

## 3.6.3  HOT FORMING

Hot forming occurs when the forming temperature is above $AC_3$. The end-forming temperature must be higher than $AR_3$ or close to this transition point. The constructor, by means of mechanical tests performed on a plate length formed by rolling and subjected to the same forming operations and the same heat treatment if necessary, must be able to demonstrate that the required mechanical properties tensile strength and fracture toughness are obtained.

*The steel mill that has delivered steel products (plate) must inform the customer of the upper and lower temperature limites to be maintained for hot forming and any heat treatment.*

*After hot forming and any heat treatment, the plate must conserve its guaranteed mechanical properties. In particular, the fracture energy on a Charpy V test specimen must remain higher than the guaranteed value for the plate steel grade before forming (see Section 3.5.2).*

## 3.7  WELDING

### 3.7.1  WELDING CONDITIONS FOR STRUCTURAL STEELS

Cold cracking is by far the most serious type of defect in the welding of structural steels, to the point where the concept of weldability of these steels is often assimilated with their sensitivity to this defect.   This is why the precautions to be observed to eliminate the risk of cold cracking occupy a predominant place in the definition of the weldability of steels for offshore structures.

The following factors are jointly responsible for cold cracking during the welding of steels:

(a) The existence of stresses in the welded joint (stresses due to restraint and shrinkage).

(b) The presence of hydrogen transferred from the melt zone to the heat affected zone (this hydrogen content depends on the filler metals and the conditions in which they are used).

(c) Quenching in the heat affected zone.

It is essential to remember that the welding behavior of steels depends closely on their composition and structure.   The welding conditions and precise values of the parameters for the adjustment of welding equipment must be determined for each type of welded joint and each steel supply.   In this respect one cannot rely on general, simple rules, or discard compliance with the qualification rules of the welding procedures.

> *The means to forestall the risks of cold cracking have led to recommendations of very unequal value and effectiveness, which are still present in certain regulations.   The qualification of welding procedures represents the most effective and most complete method to determine the weldability of a given type of joint, especially concerning the welding process and the steel quality employed.   The qualification of welding procedures is imposed by the regulations of Classification Societies [3.11] and by construction codes.*

#### A.  General recommendations

The crack testing method using implants offers quantitative data on the following points:

(a) Qualification of a welding procedure (with a given base metal, filler metal and restraint conditions).

(b) Characterization of the crack sensitivity of a base metal steel (with a given procedure, the degree of restraint and hydrogen content of filler metals being set).

The conventional method to characterize the cold cracking sensitivity of weldable steels, using different variants of the implant test (circular or helicoidal notch) is described in the following two French standards:

NF A 89-100: Cold cracking test methods using implants.

NF A 03-185: Conventional method for the characterization of the cold cracking sensitivity of weldable steels by the cracking test on helicoidal notch implants.

In the absence of a specific document for the construction of offshore structures, the recommendations recognized by the Commission d'Agrément des Aciers Soudables (Approval Commission for Weldable Steels) and compiled in the document:

NF A 36-000: Recommendations concerning the weldability of structural steel and of boiler and pressure vessel steel

can provide a framework to determine detailed specifications on the weldability of steels for offshore structures.

*From the metallurgical standpoint, a welding operation constitutes a temperature cycle whose temperature-time characteristics condition the structure and, to some extent, the composition of the constituents of the part of the base metal around the weld, known as the heat affected zone (HAZ). If the temperature cycle is such that the cooling rate of this zone exceeds a critical value, the appearance of metallurgical structures liable to hydrogen embrittlement and to cracking during cooling become inevitable.*

*The cold cracking testing method using implants offers a precise quantitative approach to the welding conditions that correspond to this critical cooling rate. The result are materialized by a so-called cracking curve, from which it is possible to determine the welding procedures that will ensure the fabrication of welded joints of satisfactory quality.*

*The welding procedure qualification test does not give information on about the consequences of restraint giving rise to lower stresses in the welded joint. This qualification is intended to ensure the observance of imposed conditions and not as an explicit guarantee that cold cracking will not occur.*

## B.  Welding of steels meeting Standards NF A 35-501 and A 36-201

For steels covered by Standards NF A 35-501 and A 36-201 except for E 460 steels, past experience has resulted in a number of alignment charts which serve to determine the following very simple:

(a) The recommended range defining all the welding conditions which, apart from specific cases, should help to eliminate the risk of cold cracking.

(b) The range corresponding to welding conditions leading to the same result if the welding operation includes preheating.

(c) The range including all the welding conditions generally advised against, but in which the welder find himself operating if he plans to proceed with both preheating and postheating in conditions to be determined with the steel producer.

The ordinate of these charts is the thickness of the members to be welded (the thickness of the chord wall for a tubular joint and the thickness at the joint for two butt welded tubes). The abscissa is the equivalent welding energy  defined by the equation:

$$E \text{ (equivalent)} = E \text{ (nominal)} . k . \text{thermal efficiency}$$

The value of the factor k depends on the type of welded joint. It is given by Table 3.2 of the commentary for some configurations commonly encountered in tubular joints. The thermal efficiency depends on the welding procedure. It is conventionally taken as 1 for welding with coated electrodes and flux welding, and 0.7 for MIG welding.

The nominal energy is given by the equation:

$$E \text{ (nominal)} = \frac{60 \ UI}{1000 \ V}$$

where E is in (kJ/cm), and U and I are the welding voltage and amperage in volts and amperes respectively, and V is the welding speed in cm/min. If the parameters U, I and V are unknown, the nominal energy may be estimated from the electrode diameter and the length of weld deposited after having consumed 10 cm of electrode. Refer to Table 3.1.

Table 3.1.

Length of weld (in cm) corresponding to 10 cm of electrode
consumed as a function of electrode diameter and nominal
welding energy; flat position (nominal electrode
efficiency: 100-115%)

| $\varnothing$ mm $E_n$ (kj/cm) | 2.5 | 3.2 | 4 | 5 | 6.3 | |
|---|---|---|---|---|---|---|
| 6 | 6.4 | 10 | 16 | 21.2 | | |
| 8 | 4.7 | 8 | 12 | 16 | 23 | |
| 10 | 3.8 | 6.5 | 9 | 12.7 | 18.3 | |
| 12 | 3 | 5.2 | 8 | 10.6 | 16.8 | |
| 14 | 2.5 | 4.5 | 6.5 | 9.0 | 14.4 | |
| 16 | — | 3.8 | 5.6 | 7.9 | 12.7 | Region of normal practice |
| 18 | — | 3.4 | 5 | 7.1 | 11.2 | |
| 20 | — | 3 | 4.4 | 6.3 | 10.1 | |
| 25 | — | 2.4 | 3.5 | 5.0 | 8.0 | |
| 30 | — | — | 2.7 | 4.3 | 6.7 | |
| 40 | — | — | 2.0 | 3.2 | 5.0 | |
| 50 | — | — | — | 2.6 | 4.0 | |

According to the data available, the charts in Figs. 3.3 to 3.9 in the commentary help to achieve the following:

(a) To find the welding energy required, by first selecting the type of electrode, the preheating conditions and the steel grade, and then, with reference to the thickness of the part to be assembled determining the minimum allowable value of the equivalent energy.

(b) To check the situation regarding the weldability range, by verifying, with reference to the thickness, the equivalent energy and the choice of electrodes, whether the operation takes place in a recommended range or whether the welding conditions have to be altered.

## Butt-welding of tubes, V joints

Table 3.2.

Coefficient k for the determination of equivalent energy

| Butt joint weld detail | α | 45° | 60° | 75° | 90° |
|---|---|---|---|---|---|
| Partial penetration V butt joint | | 0.58 | 0.60 | 0.63 | 0.67 |
| Single V butt joint | | 1.40 | 1.50 | 1.72 | 2 |
| Double V butt weld | | 0.68 | 0.75 | 0.85 | 1 |

*Remark: These coefficients are valid for thicknesses over 15 mm for the V joint and 30 mm for Y and X joints. They are conservative for lower thicknesses.*

***Tubular joints***

$$k = \frac{1}{1 + 0.5(t/T)^2} \times \frac{270°}{360° - \alpha} \quad \text{(with } \alpha \text{ in degrees)}$$

*(a) t/T = ratio of brace thickness to chord thickness.*

*(b) In this case, the angle α (see Section 2.2) varies along the intersection. Safe conditions are obtained for the minimum value of α.*

The charts given in the commentary have been prepared for non-severe restraints conditions ($\sigma < R_{eG}$). These restraint conditions probably do not guarantee certain welds against the risk of cold cracking (for example, internal stiffener weld). These charts only account for the risk of cold cracking. An upper limit of equivalent energy may also be imposed in order to satisfy requirements concerning the mechanical properties of the welded joint, and in particular, the fracture toughness of the HAZ.

### 3.7.2  PREHEATING AND POSTHEATING

The charts in Figs. 3.3 to 3.9 in the commentary serve to determine the preheating and/or postheating temperatures for a number of specific cases. Preheating must meet the following conditions: it must result from an overall or local application of heat, and should involve a width of at least five times the thickness of the members to be assembled, on either side of the joint. During multipass welding, the joint temperature, between passes, must not fall below the preheating temperature.

The following factors condition the effectiveness of postheating:

(a) Preheating and its maintenance after welding.

(b) The environment (temperature, wind, humidity).

(c) Energy of the welding passes.

(d) Frequency of execution of successive welding passes.

(e) Heat capacity of the welded members.

*Preheating and maintenance of temperatures during welding, as well as postheating, provide the means to adjust the welding temperature cycle, to ensure that the critical cooling rate is not reached. If the heat input due to welding and the form of the members are such that the critical cooling rate is always reached by "natural" cooling, the following alternatives are available:*

*(a) Preheating that increases the amount of heat to be dissipated and hence reduces the cooling rate.*

*(b) Postheating which slows down cooling, ensures the diffusion of hydrogen, and delays the formation of stress in the joint (residual stresses, stresses arising form restraint conditions).*

### 3.7.3  FRACTURE TOUGHNESS REQUIRED IN THE WELD ZONE

The weld zone consists of the zone taken up by the deposited filler metal as well as the adjacent zones called the heat affected zones (HAZ). French regulations [3.10, 3.11] and detailed specifications [3.6, 3.15] require the same minimum fracture energy in the welded zone as that imposed for the steel (base metal) (see Section 3.5.2). The temperature at which the Charpy V impact value in the welded joint (melted zone, heat affected zone) must be guaranteed is determined from the same general diagrams as those employed to determine the testing temperature on the base metal steel.

*Welding operations inherent in any construction give rise to residual stresses that may reach high levels in the welded zone. They may thus cause a deterioration in the plastic deformation properties, especially in the metal deposited and the zones affected by welding.*

*It is important to note that the ordering of steels always precedes fabrication, and that welding procedures and methods, as well as the different heat treatments subsequent to fabrication operations, are not always defined at the time of ordering. We therefore wish to draw the buyer's attention to the importance of clearly defining the specific requirements that the planned structure has to meet, at the time the steels are being ordered, and this in close liaison with the steelmaker.*

## 3.7.4  STRESS RELIEVING HEAT TREATMENT

Certain recommendations [ 3.7, 3.8 ] essentially covering structures built in the North Sea, required stress relieving heat treatment if the tube wall thickness controlling the weld dimension exceeded 50 mm for the parts of the structure in the splash zone and 63 mm for the parts of the structure in the atmospheric zone.

A slightly different approach has been proposed, based on the results of recent investigations conducted as part of ECSC's European Program. It can be summarized as follows: a stress relieving heat treatment is unnecessary provided it is demonstrated by tests (Charpy V), that the welds for dimensions greater than a certain thickness (case a: 40 mm, case b: 50 mm) display adequate fracture toughness.

Case a (Fig. 3.1):

The brace thickness joined by the weld is greater than 40 mm. This case corresponds to a weld located in a stress concentration zone.

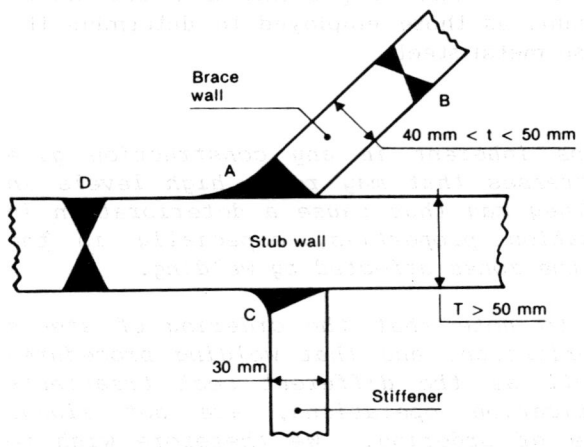

Heat treatment
required:
. Weld A
. Weld D

(when t > 50 mm,
see case b)

No heat treatment
required:
. Weld B
. Weld C

Fig. 3.1.

Case b (Fig. 3.2):

The tube thickness joined by the weld is greater than 50 mm, e.g. longitudinal weld of a welded tube and butt weld for tube butt welding (or tube/stub junction). This case corresponds to welds far from the stress concentration zones.

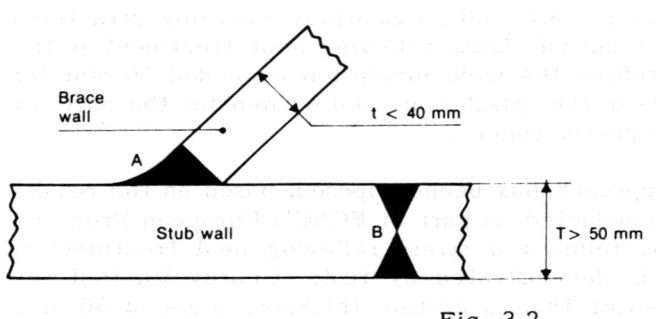

Heat treatment
required:
. Weld B

No heat treatment
required:
. Weld A

(when t > 40 mm,
see case a)

Fig. 3.2.

*Fracture toughness tests are conducted on welds executed on test specimens, in operating conditions that are absolutely identical to the procedure adopted for building the structure: the minimum impact energy level must be satisfied both in the HAZ and in the filler metal (see Part III, Section 3.5.2). For a joint, when the stress relieving of the complete member is called for, only overall treatment in the furnace is acceptable.*

*The metallurgical advantage that can be derived from stress relieving heat treatment is not always obvious. In some cases (microalloy steels with niobium or vanadium deposits with low alloy nickel electrode) the toughness may even decrease after stress relieving heat treatment.*

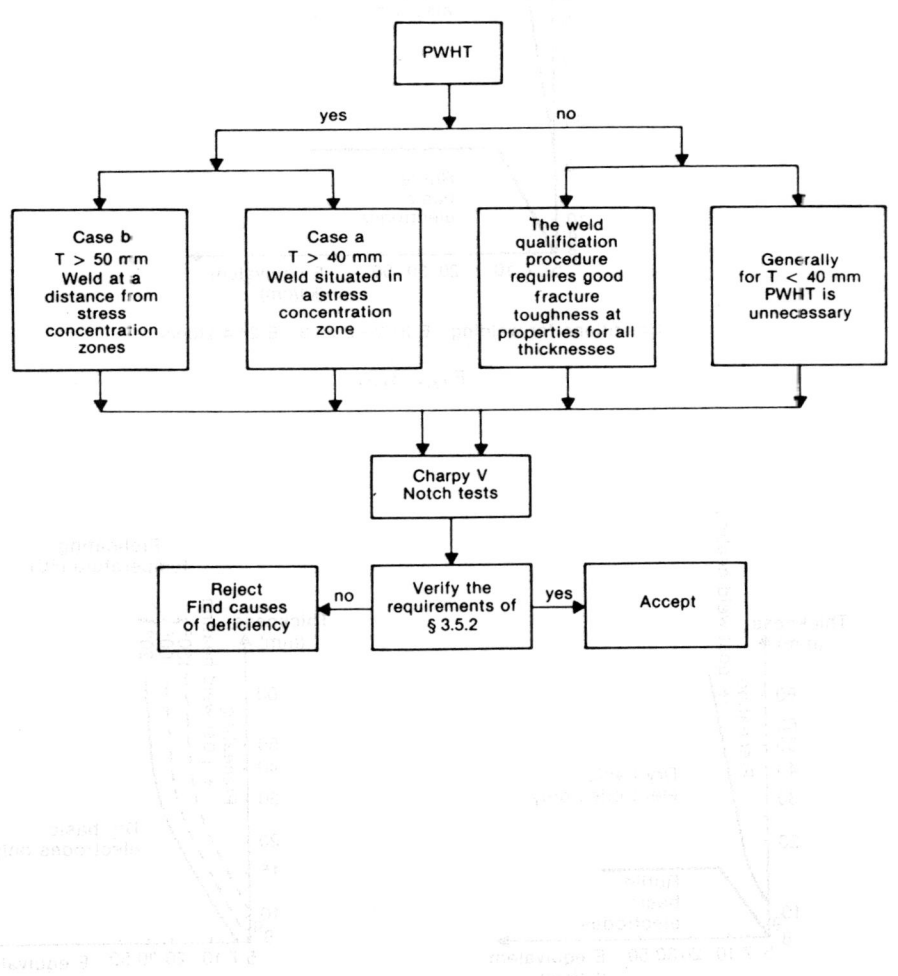

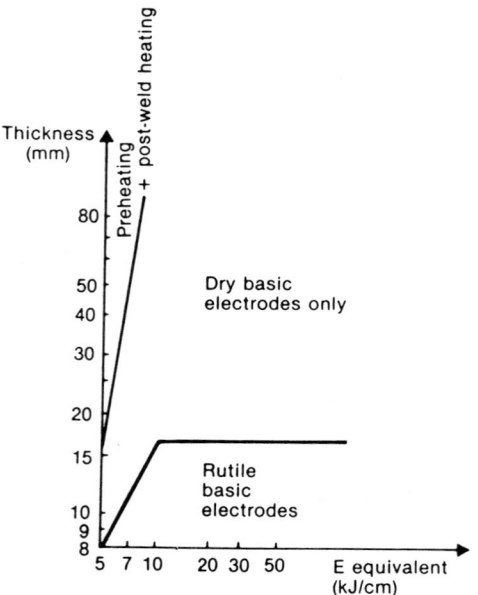

Conditions for welding - E 24-2 - E 24-3 - E 24-4 steels

Fig. 3.3.

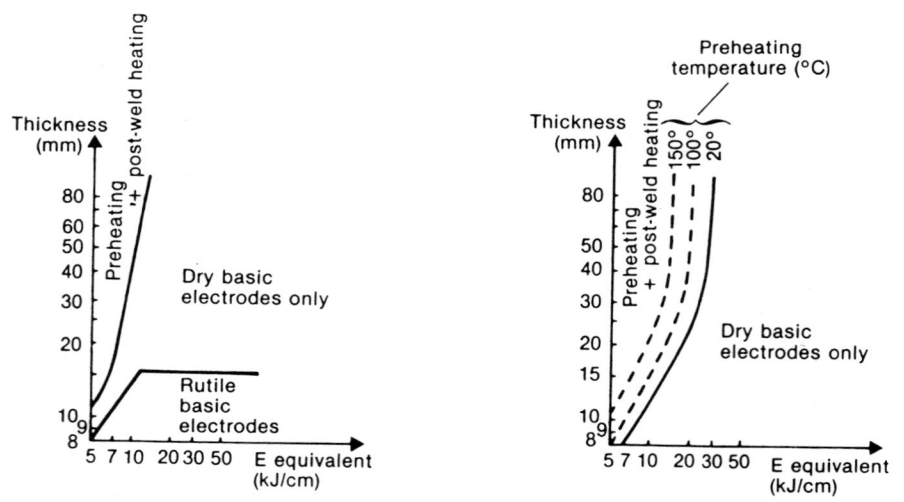

Conditions for welding - E 28-2 - E 28-3 - E 28-4 steels

Fig. 3.4.

Conditions for welding - E 36-3 - E 36-4 steels

Fig. 3.5.

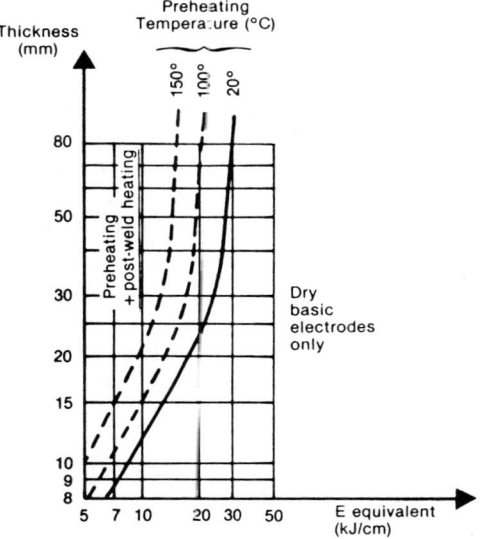

Conditions for welding - E 355 steel

Fig. 3.6.

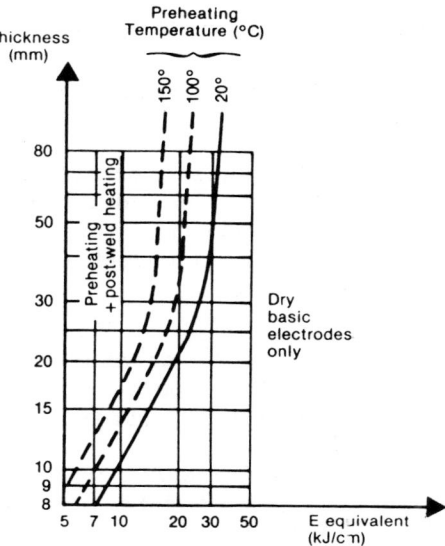

Conditions for welding - E 375 - Type 1 steel

Fig. 3.7.

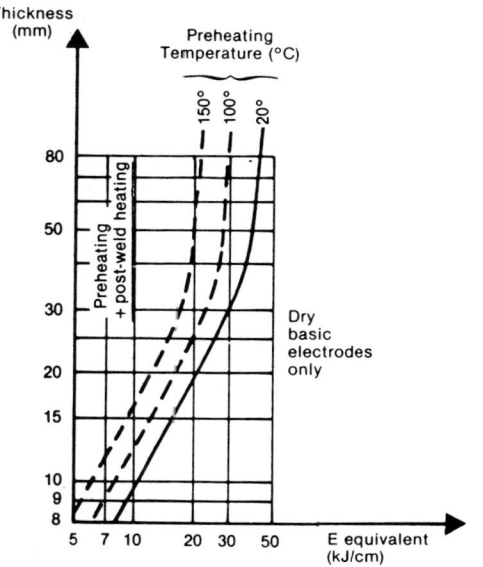

Conditions for welding - E 375 - Type II steel

Fig. 3.8.

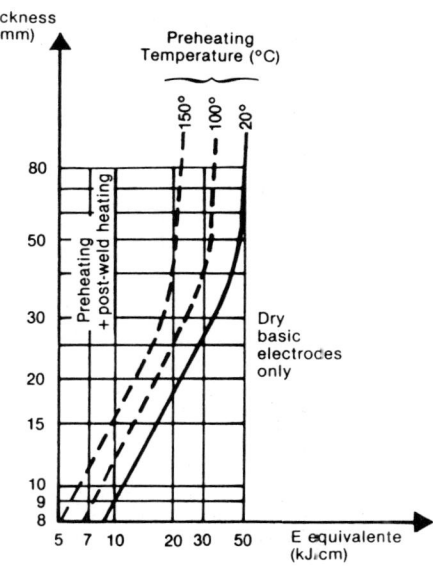

Conditions for welding - E 420 steel

Fig. 3.9.

### 3.7.5 LAMELLAR TEARING

Lamellar tearing can be avoided by imposing specific conditions on the value of reduction in area in the "short" tranverse direction (tensile test specimen taken across the thickness, or the "Z" direction), of the base product (plate or tube). As a rule, for the "special category", quality Z35 is imposed according to Standard NF A 36-202. Certain precautions may also be observed during fabrication, for example, welding onto a prior weld build up on a senstive plate.

*Due to high static or fatigue forces, tubular joints in offshore petroleum structures are often made with very thick tubes (50 to 80 mm).*

*Multipass corner joints are liable to introduce high restraint stresses, perpendicular to the plate surface, to the point of producing decohesion at the inclusions contained in the tube wall or plate (lamination defect). The risk of lamellar tearing may also be incurred under the action of high service stresses in the direction of the tube or plate thickness.*

*Previous tests, as well as past practical experience, especially in the North Sea, have shown that if the reduction in area in the short transverse direction exceeds 30%, the occurrence of lamellar tearing is extremely unlikely.*

## 3.8 STEELS FOR THE CONSTRUCTION OF TUBULAR STRUCTURES OF OFFSHORE PLATEFORMS

### 3.8.1. PLATE STEEL STANDARDS

The French standards and The Bureau Veritas regulations define a number of grades and qualities suitable for use in the construction of offshore structures. Tables 3.3 and 3.4 below give a list of these grades and qualities as a function of the imposed fracture toughness guarantees (see Section 3.5.2).

Table 3.3.

Steels with Re min. guaranteed < 300 N/mm²

| Test temperature ° C | Guaranteed Charpy-V value | | Grades | Reference standards |
|---|---|---|---|---|
| | Longitudinal direction 27 J | Transverse direction 27 J | | |
| − 40 | X | | E | Bureau Veritas |
| | X | | A 37 FP | NF A 36-205 |
| | X | X | A 48 FP | NF A 36-205 |
| − 20 | X | | E 24-4 | NF A 35-501 |
| | X | | D | Bureau Veritas |
| | X | X | A 37 FP | NF A 36-205 |
| | X | | A 37 AP | NF A 36-205 |
| | X | | E 28-4 | NF A 35-501 |
| | X | X | A 48 FP | NF A 36-205 |
| | X | X | A 48 AP | NF A 36-205 |
| 0 | X | | E 24-3 | NF A 35-501 |
| | X | | B | Bureau Veritas |
| | X | X | A 37 AP | NF A 36-205 |
| | X | | A 37 CP | NF A 36-205 |
| | X | | E 28-3 | NF A 35-501 |
| | X | X | A 48 AP | NF A 36-205 |
| | X | X | A 48 CP | NF A 36-205 |
| + 20 | X | | E 24-2 NE | NF A 35-501 |
| | X | | A | Bureau Veritas |
| | X | X | A 37 CP | NF A 36-205 |
| | X | | E 28-2 | NF A 35-501 |
| | X | X | A 48 CP | NF A 36-205 |

Table 3.4.

Steels with 300 ≤ Re min. guaranteed ≤ 420 N/mm²

| Test temperature ° C | Guaranteed Charpy-V value | | Grades | Reference standard |
|---|---|---|---|---|
| | Longitudinal direction 34 J | Transversal direction 24 J | | |
| − 40 | X | | A 52 FP | NF A 36-205 |
| | X | | E 355 FP I et FP II | NF A 36-201 |
| | X | | EH 32, EH 36 | Bureau Veritas |
| | X | | E 420 FP | NF A 36-201 |
| − 20 | X | | E 36-4 | NF A 35-501 |
| | X | X | A 52 FP | NF A 36-205 |
| | X | | A 52 AP | NF A 36-205 |
| | X | | E 355 R I et R II | NF A 36-201 |
| | X | X | E 355 FP I et FP II | NF A 36-201 |
| | X | | DH 32, DH 36 | Bureau Veritas |
| | X | | E 420 R | NF A 35-501 |
| | X | X | E 420 FP | NF A 36-201 |
| 0 | X | X | A 52 AP | NF A 36-205 |
| | X | | A 52 CP | NF A 36-205 |
| | X | X | E 355 R I et R II | NF A 36-201 |
| | X | | AH 32, AH 36 | Bureau Veritas |
| | X | X | E 420 R | NF A 36-201 |
| + 20 | X | X | A 52 CP | NF A 36-205 |

*Correspondence tables between the grades defined by French standards and the grades defined by foreign standards (DIN, BS, ASTM) and considered equivalent, are given in Annex A. These correspondences have been established on the basis of:*

*(a) Properties: Re and R.*

*(b) Guarantees of resistance against the risk of brittle fracture.*

**Remarks:**

*1. Certain steel qualities, while they do not offer a fracture toughness guarantee (e.g. ASTM A 36, ASTM A 572, Quality A of Bureau Veritas) have nevertheless been selected because the requirements concerning chemical composition should normally help satisfy minimum guarantees at + 20°C. Hence in the case of Bureau Veritas Quality A, the Charpy V tests conducted at + 20°C by steel mills that have applied for the approval of this grade have always reported fracture energies over 27 J.*

*2. For certain ASTM steel qualities, when ordering, to make sure that the correspondence is valid, it is necessary to add supplementary requirements concerning fracture toughness.*

*3. Certain steel qualities with guaranteed fracture toughness at a given temperature t have been treated as qualities with guarantees at a temperature t - 20°C, due to the high energy values fixed for temperature t. In fact, the following Energy-Temperature equivalences are acceptable:*

*48 J at 0°C or 40 J at - 20°C or 27 J at - 40°C*
*56 J at 0°C or 48 J at - 20°C or 40 J at - 40°C*
*or 27 J at - 50°C*

*However, these equivalences can only be taken into consideration if the guaranteed energies are given at temperatures of at least - 20°C.*

*4. Certain steel qualities, ASTM in particular, exhibit high maximum C contents (e.g. ASTM A 36, A 573, A 709, etc.) with values over 0.20% in particular. It is up to the builder to limit these contents by supplementary requirements, should he feel this necessary.*

## 3.8.2  STANDARDS CONCERNING STEEL TUBES

No specific standard exists covering the precise aspect of the use of steel tubes for offshore petroleum structures. Standards are mentioned in the comments and analyzed in Annex B, for several grades and qualities of steel tubes available on the French market and offering adequate guarantees. Supplementary requirements related to grades and qualities, fabrication, state of delivery, inspection, packaging and acceptance must be specified to the manufacturer.

*The following NF standards are analyzed in Annex B:*

*NF A 49-211: Steel tubes, seamless plain-end carbon steel tubes for the transport of fluids at elevated temperatures, dimensions. Technical delivery conditions. Amended by: Erratum December 1981. (June 1981)*

NF A 49-213: *Steel tubes. Seamless unalloyed and MO or CR-MO alloy steel tubes for use at high temperatures. Dimensions (with normal tolerances). Technical delivery conditions (Nov. 1983).*

NF A 49-240: *Steel tube longitudinally buttwelded plain end for pressure vessels and pipe systems used at low temperatures. Dimensions. Technical conditions of delivery (Sep. 1983).*

NF A 49-253: *Steel tubes. Longitudinally fusion welded non alloy steel and ferritic alloy steel tubes for use at elevated temperatures. Dimensions. Technical Delivery conditions (Sep. 1982).*

NF A 49-400: *Steel tubes. Longitudinal electric resistance welded unalloyed steel tubes. 17.2 inferior or equal to D inferior or equal to 406.4 mm for the transport of pressurized fluids. Dimensions. Technical delivery conditions (March 1982).*

NF A 49-401: *Steel tubes. Longitudinally fusion welded unalloyed steel tubes for pipes and pressure vessels. Dimensions. Technical delivery conditions (Dec. 1981).*

NF A 49-410: *Steel tubes. Seamless plain-end carbon steel tubes for the transport of pressurized fluids. Dimensions. Technical delivery conditions. Amended by: Erratum December 1981 (June 1981).*

NF A 49-411: *Steel tubes. Seamless high performance tubes. 60.3 mm D 406.4 mm of unalloyed steels for butt-welding pressurized fluid transport pipelines. Dimensions. Technical conditions of delivery (May 1982).*

NF A 49-501: *Steel tubes. Hot finished structural hollow sections. Dimensions. Technical delivery conditions (Jan. 1982)*

NF A 49-541: *Steel tubes. Cold finished structural hollow sections. Dimensions. Technical delivery conditions (April 1983).*

## REFERENCES

3.1     Bases de Choix des Aciers en Construction Métallique, Vol. 1, OTUA, 1970

3.2     Sanz, G., La Rupture des Aciers, Vol. 1, La Rupture Fragile, "Les propriétés d'emploi des aciers", IRSID/OTUA Collection, September 1974.

3.3     Granjon, H., La fissuration à froid en soudage d'aciers (Doc. IIS/IIW 384-71), Soudage et Techniques Connexes, Nos 3/4, 1972.

3.4     Rousseau, P., Les aciers utilisés en construction métallique, AFNOR/BNS Collection, published by AFNOR, 1st Edition 1977.

3.5     Thomas, J.M. and Rousseau, P., Ecrouissage et vieillissement des tôles ou des aciers, Technical Bulletin of the Bureau Véritas, December 1982.

3.6     Matériaux pour structures marines en acier, General Specifications for Structures No. 211, TOTAL, SP-STR-211.

3.7     Rules for the Design, Construction and Inspection of Offshore Structures, Det Norske Veritas, 1977.

3.8     Code of Practice for Fixed Offshore Structures, British Standards Institution, BS 6235: 1982.

3.9     Sanz, G., Risque de rupture fragile, Essai de mise au point d'une méthode quantitative de choix des qualités d'aciers vis-à-vis du risque de rupture fragile, AFNOR/IRSID, 1981.

3.10    Rules and Regulations for the Constructor and Classification of Steel Ships, Offshore Platforms and Semi-Submersibles Characteristics and Control of Materials, Bureau Véritas, Paris, 1980 (with amendments and additions, January 1982).

3.11    Rules and Regulations for the Constructor and Classification of Offshore Units, Bureau Véritas, Paris 1975 (with amendments and additions, No. 1, August 1982).

3.12    Courbes dureté/paramètres de refroidissement en conditions de soudage, IRSID Collection, IRSID publication 1977.

3.13    Conseils pour le soudage des aciers de construction métallique et chaudronnée à la limite d'élasticité garantie 420 N/mm$^2$, ATS/OTUA Collection, 1980.

3.14 Yurioka et al, Determination of necessary preheating temperature in steel welding, 63rd Annual Convention, AWS, April 1982.

3.15 Liégeois, J., Considérations pratiques sur le soudage et la soudabilité des aciers microalliés à haute limite d'élasticité, Soudage et Technique Connexes, Vol. 34, Nos. 9/10, 1980.

3.16 Debiez, S., Synthèse d'un ensemble de résultats d'essais de fissuration et application à la détermination pratique des conditions de soudage des aciers du type E36, Soudage et Techniques Connexes, Vol. 34, Nos 9/10, 1980.

3.17 De Soras, D. and Charleux, J., Ecrouissage et vieillissement des tôles en acier suite à un formage à froid par pliage ou par roulage suivi d'un effet thermique dû au soudage, Technical Bulletin of the Bureau Véritas, December 1982.

3.18 Travaux collectifs sur l'arrachement lamellaire, Quelques résultats et commentaires, Soudage et Techniques Connexes, March/April 1977.

CHAPTER **4**

# Corrosion Protection

## 4.1 CORROSION AND FATIGUE

This Section is not intended to provide a detailed analysis of the effect of corrosion on the fatigue strength of welded tube assemblies, but to describe briefly the protection techniques routinely employed to prevent steel corrosion in offshore plateforms. The effect of corrosion and of cathodic protection is examined in Section 6.2, Part III, which deals with the changes in the reference S-N curve under the effect of these factors.

It must be kept in mind that effective corrosion protection of an offshore structure automatically considerably limits the harmful effects on the behavior of the structure due to fatigue. The basic principle of preventing corrosion-fatigue is to employ the means that are generally used against corrosion.

Like all structures exposed to the marine environment, offshore structures are subjected to substantial attack of marine corrosion. The effects of corrosion, which are themselves serious, also play a non-negligeable role on the fatigue behavior of welded tubular joints. The term corrosion fatigue embodies fatigue processes which take place in the corrosive environment. The damage observed in corrosion-fatigue is often greater than the sum of the damages caused by corrosion and by fatigue considered separately.

In fact, it is impossible to speak of a "fatigue threshold" below which variations in cyclic stresses are of an insufficient magnitude to initiate cracking. Fortunately, the use of effective cathodic protection in the submerged part of the structures, usually helps to ensure that their fatigue behavior will be similar to that "in air".

## 4.2 MARINE CORROSION

Expressed as a uniform loss of steel, the seawater corrosion rate may seem relatively acceptable. Corrosion rates vary according to the exposure zone. They generally lie between 0.1 and 0.2 mm/year in the initial years, and then decline to values as low as 0.05 mm/year. In actual fact, corrosion is not uniform, and local attack related to heterogeneities at the steel-seawater interface may imperil a structure by perforation or by an excessive loss of the load resisting cross-section.

1) In a buried zone, corrosion is generally lower because of limited access to oxygen. However, specific processes occur, particularly in mud, due to the possible development of sulfate-reducing bacteria in anaerobic medium, which may lead to bacterial corrosion craters.

2) In the atmospheric part of steel structures, corrosion may be high locally. Hence the most corroded zone is the one situated just at the wave splash zone. This area is affected by a combination of the constant substantial renewal of oxygen in the seawater film that periodically covers the steel, as well as the mechanical and thermal effects of the oxides formed. This joint action leads to high corrosion rates (about 0.5 mm/year at ambient temperature, and up to 5 or 10 mm/year on hot risers).

3) In the part of the structure that is totally and permanently exposed to the atmosphere (atmospheric zone), spray or condensations of salt water due to the moisture in the air cause rapid rusting of the steel.

4) The reader's attention is drawn to potential and specific problems, such as a high corrosion rate, related to specific environments. These special cases (excessive speed of marine currents, presence of sulfide, etc.) mean that it is recommended to closely identify the parameters of the marine environment to which the structure is exposed. The knowledge of these parameters helps to select and optimize the protection systems.

*The marine environment is highly corrosive to many metallic materials, especially the carbon-manganese steels employed for the construction of offshore structures. Seawater is a highly conductive electrolyte (resistivity about 25 Ω cm), generally saturated with dissolved oxygen in equilibrium with air (about 7%) and highly charged with chloride ions (about 19%). This combination of factors favours steel corrosion, which occurs by an electrochemical process whose main elementary reactions are:*

*(a) Oxidation of the steel (anodic process):*

$$Fe \longrightarrow Fe^{++} + 2e^{-}$$

*(b) Reduction of oxyden (cathodic process):*

$$\frac{1}{2} O_2 + H_2O + 2e^{-} \longrightarrow 2\ OH^{-}$$

*Chloride ions play a complex role. In addition to the ionic conductivity that they create, their high absorption capacity favours corrosion and reduces the protective power of complex iron oxides and hydroxides (rust) which are formed by secondary chemical reactions between the chemical species created by the elementary electrochemical reactions.*

## 4.3 PROTECTIVE SYSTEMS

### A. Cathodic protection

The primary means employed to prevent the corrosion of structures above sea level, which exploits the electrochemical character of corrosion by seawater, consists in lowering the electrochemical potential of the steel below a value at which the anodic reaction of iron dissolution is practically eliminated, in favor of the cathodic reaction of the reduction of dissolved oxygen. This process is called cathodic protection. Cathodic protection is not recommended for high strength steels.

Offshore platforms are always provided with cathodic protection. The impressed current system is used fairly often on mobile units, but less frequently on stationary platforms, especially in rough seas for which problems of system robustness arise. As a rule, platinized titanium or platinized niobium is used for overflow impressed current anodes. More frequently, stationary platforms are protected by sacrificial anodes using aluminium alloys activated with indium or mercury to prevent passivation.

In rare cases in which the entire structure is coated, zinc alloys are often preferable because they are considered more reliable in case of low stresses.

In both cases, the potential of the structure cannot fall below that of the sacrificial anodes, or about - 1.050 V/Ag - Ag Cl, preventing any excess of cathodic polarization and hence any liberation of hydrogen (and the potential harmful effects that this may cause on certain steels, including that of fatigue behavior).

With an impressed current system, however, it is necessary to monitor the anode depletion rate to eliminate this risk.

### B. Coatings

In addition to active protection methods, the use of miscellaneous coatings, which isolate the steel from the corrosive environment is fairly widespread for offshore structures. The atmospheric surfaces of the structures are protected by systems of marine paints about 250 µ thick. Apart from special systems employed for specific purposes (high temperature, non-skid coating, etc.) the most widespread system involves a zinc ethylsilicate primer covered by an epoxy resin seal coat after preparation by a bonding coat. The finish uses acrylic or polyurethane base materials. Some items, such as ladders, railings and gratings, are galvanized.

Systems consisting of a zinc ethylsilicate primer covered with two or three coats of epoxy pitch and possibly with one anti-fouling finish, are generally applied to protect the splash zone. Thicker coatings (epoxy with filler, solvent-fray polyurethane) are currently being investigated. Cladding by "Monel" plate is sometimes also employed. In the submerged zone, paint systems using two coats of epoxy pitch are employed, either for the entire surface, or more usually to coat specific zones such as the part in contact with mud (to prevent bacterial corrosion), and increasingly, certain parts that are relatively inaccessible to the cathodic protection current (joints, conductor pipes, etc.).

*In seawater, the natural potential of steel is around -0.650 V in relation to the silver-chloride electrode generally used. If it is raised to a more negative value than -0.800 V, it is protected cathodically, i.e. its surface functions exclusively as a cathode under the effect of a direct electric current generated by seawater. This current is dispersed in the electrolyte by means of anodes installed on the structure, either naturally by simple electrical connection of the anodes to the structural frame, or by means of a d.c. source installed at the surface.*

*The former case corresponds to a galvanic system employing "sacrificial" anodes, consisting of a preferentially consumed material that is less noble than the steel. The second case corresponds to an impressed current system, whose "overflow" anodes must be as inert as possible and insulated electrically from the structural frame.*

# REFERENCES

4.1 Mackdanz, C.H.., Protection of offshore structures, Materials Protection, 4 (10) 83, 1965.

4.2 Hanson, H.R. and Hurst, D.C., Corrosion control, Offshore platforms, OTC 1969, Paper No. 1042.

4.3 Baribault, J.D., Cathodic protection for offshore structures, Oil and Gas Journal, p. 91, April 1963.

4.4 Burgbacher, J.A., Cathodic protection of offshore structures, Materials Protection, 7 (4) 26, 1968.

4.5 Potosnak, C.S., Tate, R.E. and Talbot, C.J.H., Cathodic protection of offshore structures, Offshore Europe, p. 328, 1968.

4.6 Heuze, B., Protection cathodique d'engins flottants et d'ouvrages maritimes, Construction, 23 (9), 1968 and 24 (10), 1969.

4.7 Lehmann, J.A., Cathodic protection of offshore structures, OTC 1969, Papier No. 1041.

4.8 Compton, K.G. and Lee Craig, J.H., Cathodic protection of offshore structures, OTC 1970, Paper No. 1271.

4.9 Grosz, O.L., Cathodic protection for platforms, Oil and Gas Journal, 17 November 1969.

4.10 Davis, J.G., Doremus, G.L. and Graham, F.W., The influence of environmental conditions on the design of cathodic protection systems for marine structures, OTC 1971, Paper No. 1.

4.11 Fitzgerald, J.H., Cathodic protection of stationary marine structures, Materials Protection and Performance, 11 (5) 23, 1972.

4.12 Mackay, W.B., North Sea offshore cathodic protection, OTC 1974, Paper No. 1957.

4.13 Compton, K.G., Lee Craig, Jr.,H. and Smith, C.A., Considerations of importance in the cathodic protection of marine structures, Corrosion NACE 74, Paper No. 85.

4.14 Thome, O. and Hansen, A.H., Corrosion problems on offshore structures, Northern Offshore, (1) 48, 1957.

4.15   Jensen, F.O., Corrosion and protection of offshore steel structures, Corrosion NACE 76, Paper No. 182.

4.16   Roche, M. and Samaran, J.P., Specificité de la protection cathodique offshore, Pétrole et Techniques, No. 289, p. 35, March 1982.


**Works:**

4.17   Roche, M., Protection contre la corrosion des ouvrages maritimes pétroliers, IFP Publication, Editions Technip, 1978.

4.18   Offshore Cathodic Protection, NACE 1975, Order No. 52088.

4.19   Fink, F.W. and Boyd, W.F., The Corrosion of Metals in Marine Environments, Defense Metals Information Centre, Report 245, Battelle, 1970.

4.20   Rogers, T.H., Marine Corrosion, G. Newnes Ltd., London, 1968.

4.21   Laque, F.L., Marine Corrosion, Causes and Prevention, J. Wiley and Sons, New York and London, 1975.

4.22   Collee, R., Corrosion Marine, CEBEDOC, Liège, 1975, distributed by Eyrolles, Paris.

4.23   Corrosion et Protection Offshore, International Symposium, CEFRACOR, Paris, 7/11 May 1979.

4.24   Lemoine, L. and Thébault, J., Corrosion Marine, Bibliographie, CNEXO Scientific and technical Report No. 36, 1977.

4.25   Control of Corrosion of Steel Fixed Offshore Platforms Associated with Petroleum Protection, NACE Standard RP-01-76.

4.26   Corrosion Marine, Moyens de Protection, International Symposium, CEFRACOR, La Baule, 4/7 June 1974, published by CNEXO, Actes de Colloques No. 3, 1974.

# PART II

# ANALYSIS
# OF THE STATIC STRENGTH
# OF TUBULAR JOINTS

# CHAPTER 1

# Introduction

The analysis of the static strength of welded tubular joints entails:

(a) The determination or forecasting of maximum loads that will be exerted on a given joint during the life of the structure.

(b) The existence of ultimate static strength formulas that serve to confirm that these maximum loads will not cause the failure of the joint.

## 1.1  LOAD CALCULATIONS

The determination or forecasting of maximum loads requires the following:

(a) Consideration of all loading categories (dead weight, operating loads, environmental loads, etc.).

(b) Definition of the loading cases that need verification. These loading cases may concern normal conditions or specific conditions due to the extreme values that may be assumed by the different climatic and oceanographic variables. They also concern the temporary conditions encountered during the preliminary phases in the life of the structure (fabrication, launching, towing).

(c) A model to calculate the loads from the forces. The different models are described in Part II, Section 3. Based on conventional structural analysis, they furnish the forces acting on the joint. The loads taken into account are the axial force $N_x$, the bending moment $M_y$ and the bending moment $M_z$ (Fig. 1.1). The other loads are ignored.

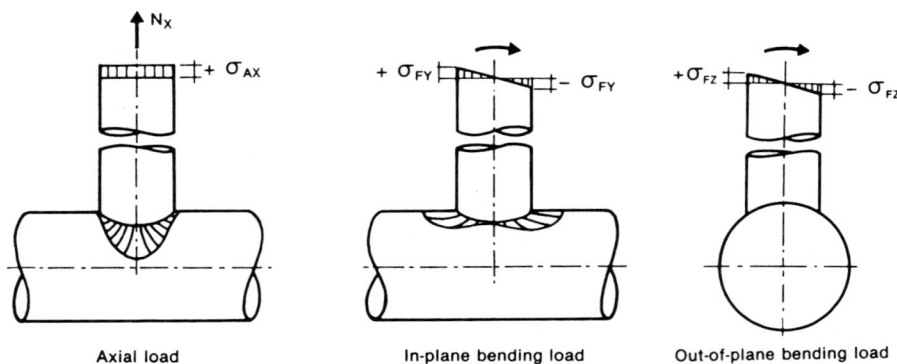

Axial load          In-plane bending load          Out-of-plane bending load

Fig. 1.1.  Loads taken into account in static strength formulas.

## 1.2  STATIC STRENGTH FORMULAS

Ultimate static strength formulas are based on the statistical treatment of test results. The tests are conducted on simple joints (X, T, Y, N and K) and on N and K joints with overlap. In all cases, a single type of loading is applied (axial load, in-plane bending, out-of-plane bending) and the boundary conditions are relatively simple.

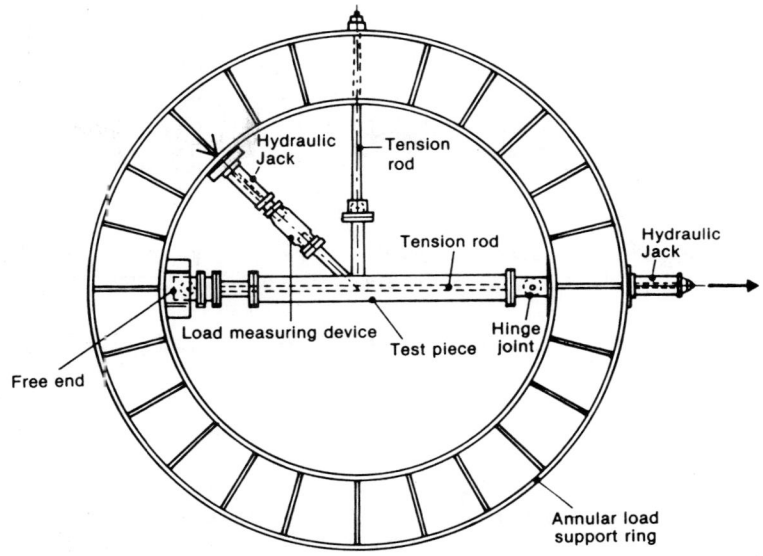

Fig. 1.2. Experimental rig (N joint).

It must be kept in mind that a welded tubular joint of a real structure does not normally conform to the conditions stated above. What is involved is a combination of load as well as far more complex boundary conditions. The geometry itself may deviate from the standard case analyzed as is the case of a joint with several brace connections or a stiffened joint. These considerations are illustrated by Fig. 1.2, which shows a schematic diagram of an experimental test rig for an N joint.

*Tubular joints display a static strength capacity far beyond the load that produces initial plastification in the assembly. The ratio of the ultimate load causing the collapse of the assembly to the "elastic load" may vary from 2.5 to 8 [1.1]. In the present state of knowledge, a theoretical (analytical) elastic-plastic or numerical approach is not adequate as a basis for regulations or formulas. In fact, these approaches are not yet sufficiently developed and tests have not yet been conceived with a view to recording the measurement corresponding to the collapse criteria adopted in these approaches. Hence they cannot serve to validate a numerical model.*

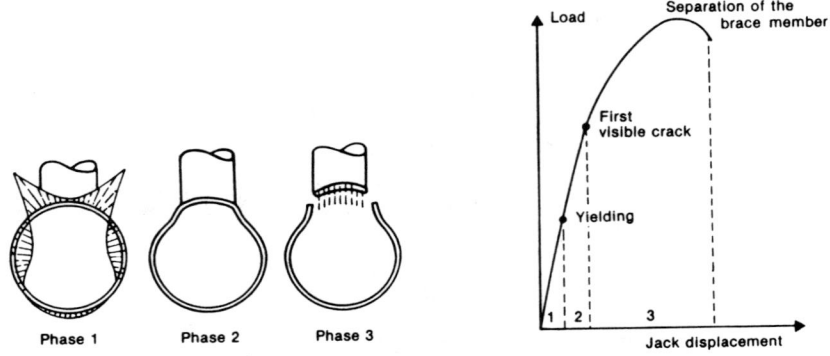

*Fig. 1.3. T joint under tensile loading.*

The experimental method makes the following assumptions:

(a) The definition of test collapse criteria, corresponding to a given instant either in phase 2, or in phase 3 (Fig. 1.3).

(b) Measurement of the ultimate load corresponding to this criteria.

(c) Statistical treatment of the test results, in which the test ultimate load ($P_u$) is "explained" by the variables or geometric parameters of the joint and by the mechanical properties of the material employed:

$$P_u = f(D,T,t,L,\alpha,\beta,\gamma,\tau,\Theta,\sigma_y,\frac{\sigma_y}{\sigma_u})$$

**REFERENCES**

1.1 Marshall, P.W. and Toprac, A.A.M., Basis for tubular joint design, Welding Journal, Welding Research Supplement, pp. 192-201, May 1974.

1.2 Marshall, P.W., Basic considerations for tubular joint design in offshore construction, WRC Bulletin, No. 193, April 1974.

# CHAPTER 2

# Determination of Extreme Values of Climatic and Oceanographic Parameters

## 2.1 WAVE

### 2.1.1 DEFINITION

The so-called "project design wave," i.e. the extreme wave used in the loading cases (see Chapter 3) which are the subject of verification, is generally a value presented as deterministic (for example, so-called fifty-year or one hundred-year wave). By contrast, the approach recommended here presents the crest-to-trough extreme value over a period of T years $(H_{ext(T)})$ as a random variable of which the probability distribution must be determined or estimated.

The extreme parameters relative to wave action are hence the mode or the estimation of the mode of $H_{ext(T)}$ with which a confidence interval is associated corresponding to a given probability (Fig. 2.1), a range of associated periods, as well as a propagation direction.

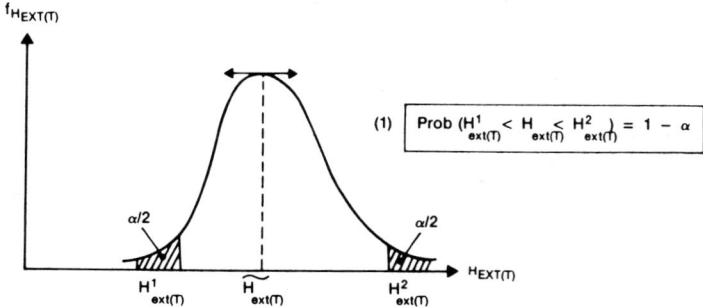

Fig. 2.1. Distribution of the crest-to-through extreme value H over T years - $H_{ext(T)}$.

This procedure is the only one which provided a valid clear basis for the risk analysis that underlies the analyses normally carried out. Several methods exist to determine the extreme parameters of crest-to-trough height H. They all refer to the theory of extreme values. We shall therefore begin by discussing the first result of this theory (see Section 2.1.2) and then go on to the existing methodologies (see Section 2.1.3). Section 2.1.4 states the recommendations of this guide, while Section 2.1.5 focuses on the use of available data. The question of the range of periods associated with the extreme parameters of the crest-to-trough height H is discussed in Section 2.1.6.

## 2.1.2 THEORY OF EXTREME VALUES

Let us consider a random variable X, with a continuous distribution function $F(x)$, and assume that $X_1$, $X_2$,...,$X_n$ is a series of small n independent observations deriving from the parent random variable X. The theory of extremes analyzes the behavior of values associated with the extremes of $X_1$,...,$X_n$ when the sample size n increases indefinitely. The method deals with the limit properties of the maximum probability distribution of the sample $X_{ext(n)} = \max (X_1,...X_n)$.

Due to the two central assumptions of independence and equi-distributivity,

$$F_{X_{ext(n)}}(x) = Prob(X_{ext(n)} < x) = Prob(\forall i \in \{1,2,...n\}/X_i < x)$$

$$= [F(x)]^n \qquad (1)$$

the following theorem can therefore be demonstrated. Series $a_n$ and $b_n$ ($a_n$ positive) exists such that:

$$\lim_{n \to \infty} \left[F(a_n x + b_n)\right]^n = \lim_{n \to \infty} Prob(X_{ext} < a_n x + b_n) = G(x) \qquad (2)$$

where $G(x)$ is one of the three following distribution laws:

Gumbel's Law (or type I):

$$\Lambda(x) = e^{-e^{-x}} \qquad (-\infty < x < +\infty) \qquad (3.1)$$

Frechet's Law (or type II):

$$\Phi_a(x) = e^{-(x)^{-a}} \qquad \text{if } x > 0, \quad 0 \text{ if } x \leq 0 \qquad (a > 0) \qquad (3.2)$$

Weibull's Law (or type III):

$$\Psi_a(x) = e^{-(-x)^a} \quad \text{if } x < 0, \quad 1 \text{ if } x \geqq 0 \quad (a > 0) \quad (3.3)$$

Hence this theorem does not confirm that $X_{ext(n)} = \max(X_1, \dots X_n)$ obeys a distribution law of extreme values when n tends towards infinity, but that it does so for a certain function

$$\frac{X_{ext(n)} - b_n}{a_n}$$

of $X_{ext(n)}$.

The passage to the distribution law of $X_{ext(n)}$ is nevertheless immediate since:

$$\left[F(a_n x + b_n)\right]^n \simeq G(x) \Longrightarrow \left[F(x)\right]^n \simeq G\left(\frac{x - b_n}{a_n}\right) \quad (4)$$

The problem finally reduces to determining the type of limit distribution (I, II or III) and the expressions giving $a_n$ and $b_n$. The foregoing statement does not mean that the series $a_n$ and $b_n$ are convergent. Hence it is the expression of these series as a function of n that is used in formula (4). Since this is used with fixed n (100 or 500 for example), it is also necessary to specify the error committed when

$$\left[F(a_n x + b_n)\right]^n$$

is replaced by its limit distribution (see below):

1. *The choice of constants $a_n$ and $b_n$ is not unique. If, for example, it is necessary for*

$$\lim_{n \to \infty} Prob(\frac{X_{ext(n)} - b_n}{a_n} < x) = \lim_{n \to \infty} Prob(X_{ext(n)} < a_n x + b_n) =$$

$$= \lim_{n \to \infty} \left[F_X(a_n x + b_n)\right]^n = \Lambda(x) = e^{-e^{-x}}$$

*it suffices to use the following expressions for $a_n$ and $b_n$ when the type of limit distribution of $X_{ext(n)}$ is effectively of type I ($\Lambda$):*

$$b_n = F^{-1}(1 - \frac{1}{n})$$

$$a_n = F^{-1}(1 - \frac{1}{ne}) - F^{-1}(1 - \frac{1}{n})$$

When the long-terme modelling of $H_{1/3}$ is given by a Weibull law with two parameters ($F(x) = 1-exp\{-[\frac{1}{3.52}]^{1.43}\}$), $a_n$ and $b_n$ are given by:

$$b_n = 3.52[\ln(n)]^{1/1.43}$$

$$a_n = 3.52[[\ln(ne)]^{1/1.43} - [\ln(n)]^{1/1.43}]$$

For a value of n equal to 100, the direct and approximate calculations give the following results with respect to the probability that the extreme value observed in 100 repetitions is less than 12:

(a) Direct calculation:

$$F_{X_{ext(100)}}(12) = [F_X(12)]^{100} = \left[1 - e^{-\frac{12}{3.52}^{1.43}}\right]^{100} =$$

$$= 0.733$$

(b) Approximate calculate using the theorem on the existence of an extreme limit distribution for $|F(a_n x + b_n)|^n$:

$$\left.\begin{array}{l} b_{100} = 10.24 \\[2em] a_{100} = 1.51 \end{array}\right\} \Longrightarrow \frac{12 - b_{100}}{a_{100}} = 1.17 \quad \text{and}$$

$$F_{X_{ext(100)}}(12) \simeq G\left(\frac{12 - b_{100}}{a_{100}}\right) = e^{-e^{-1.17}} = 0.733$$

2. The Von Mises and Jenkinson criteria help to characterize the areas of attraction corresponding to the

asymptotic laws of type I, II, and III.  On the whole, it may be stated that the type of limit distribution depends on the behavior for large x values of the probability density $f'(x) = \dfrac{dF(x)}{dx}$ of the parent distribution.

The concept of convergence rate also serves to assess the error committed, for a given n, if $F^n$ is replaced by its limit distribution.  For example, note the following results [2.1.3]:

(a) If F is a Guassian distribution, the approximation of $F^n$ by G (Gumbel) is of the order ln(n).

(b) If F is an exponential distribution, the approximation of $F^n$ by G (Gumbel) is of the order 1/n.

(c) If F is a Pareto distribution, the approximation of $F^n$ is of the ordre $1/(n^c)$ where c is greater than 1.

## 2.1.3 CHOOSING A METHODOLOGY

Two guidelines exist concerning the determination of a probability distribution of an extreme value relating to the wave environment parameter.

A. The first guideline consists in estimating the long-term model of H, the crest-to-trough height, and then taking as the extreme design project parameter the mode of the distribution of $H^n$ for a value of n corresponding to a given time interval (20, 50 or 100 years).  This approach corresponds to the direct calculation of $[F(x)]^n$.

B. The second guideline consists of selecting the highest observed values of H by choosing a threshold whose influence on the extreme design parameters finally selected must be clarified.  This approach employs the theoretical results recalled in Section 2.1.2 (evaluation of asymptotic laws of types I, II and III).

Approach B actually combines two methods - modelling and distribution tails - which are illustrated in diagram 1 (Section 2.1.4).  The comments of that Section contain a detailed example of approach A, as well as a number of remarks concerning asymptotic methods.  These serve to justify the recommendations in Section 2.1.4.

It remains necessary to:

(a) Stipulate the theories or theoretical elements that are employed, with an aim towards verifying the hypotheses underlying them.

(b) With respect to data, it is necessary to specify the basic process analyzed (individual waves, maximum storm waves, maximum annual waves) and the procedures for making up the representative sample of maximum values of the basic process, if it does not involve the individual waves.

Added to this, is the problem of the quality of the data compiled, and hence, in case of imperfection of the data acquisition system, the possible need to resort to error detection and measurement correction techniques.   In approach B, measurement correction can be reflected directly on the extreme parameters selected, and hence without having to repeat the entire computation procedure (Table 2.1).

Table 2.1.

Characterization of two approaches for the determination of extreme project design wave parameters relative to crest-to-trough height H

|  | Theoretical elements | Data employed |
|---|---|---|
| Approach A | Assessment of the mode of $F(x)^n$ | In general, all the individual waves |
| Approach B | Assessment of asymptotic law (I, II or III) | Distribution tails |

### *An illustration of approach **A**: the **DnV** regulation.*

*The construction of the long-term model of H is achieved by combining the long-term distribution law of the characteristic parameters of a short-term sea state with the H distribution of a given state of the sea:*

$$Prob(H \leq H^*) =$$

$$\int_0^\infty \int_0^\infty \int_0^{H^*} P_{H/(\overline{H}_{1/3}, \overline{T}_\eta)} (z/(x,y)) f_{\overline{H}_{1/3}, \overline{T}_\eta} (x,y) \, dx \, dy \, dz \tag{5}$$

*Using a number of assumptions, concerning the short-term H distribution (Rayleigh's law, i.e. assuming that the elevation of the free surface is modelled by a Gaussian stationary and ergodic process and that the spectrum has a narrow band $\varepsilon = 0$), the probability defined by (5) can be written:*

$$F_H(H^*) = Prob(H \leq H^*) = 1 - exp\left[\left(\frac{H^*}{C(c,d,\gamma)H_c^{1/d}}\right)^{D(\gamma,d)}\right] \quad (6)$$

It consists of a two parameter Weibull distribution.

Equation (6) is the one found in [2.4, 2.5], c and d are the two parameters of the visual height $(H_V)$ to significant height $(H_{1/3})$ relation while $H_c$ and D are the two parameters of the Weibull law modelling the long-term distribution of the visual height $H_V$.

Assuming the formal framework of the theory of extreme values defined in Section 2.1.2, the series $X_1, \ldots\ldots X_n$ of n observations in this case is that of n waves occurring during a given interval T. This is actually an estimation of this number of waves. Using Equation (1), the following equations are obtained:

$$F_{H_{ext(n)}}(H^*) = Prob(H_{ext(n)} < H^*) = [F_H(H^*)]^n \quad (7)$$

$$f_{H_{ext(n)}}(H^*) = \frac{d}{dH^*}\left(F_{H_{ext(n)}}(H^*)\right) = n\,F_H(H^*)^{n-1}f_H(H^*) \quad (8)$$

Mode $\tilde{H}_{ext(n)}$ is defined by $\frac{d}{dH^*}(f_{H_{ext(n)}}(H^*)) = 0$ and is hence a solution of:

$$F_H(H^*)\,\frac{d}{dH^*}\left(f_H(H^*)\right) + f_H^2(H^*)\,(n-1) = 0 \quad (9)$$

The solution of Equation (9) is generally difficult to obtain analytically. However, for a parent distribution whose tail is of the exponential type, one can use the asymptotic property of the mode $\tilde{X}_{ext(n)}$ of the law $(X_{ext(n)}, F(x)^n)$. This property is expressed as follows:

$$\tilde{X}_{ext(n)} \simeq X_e \quad where \quad F(X_e) = 1 - \frac{1}{n} \quad (large\ n)$$

$$(10)$$

The resolution of (9) for a high value of n reduces to using the property (10) for the long-term distribution function of H defined in (6) (Weibull's law, hence an exponential tail).

*This gives:*

$$\tilde{H}_{ext(n)} = C\, H_c^{1/d} [\ln(n)]^{1/D} \qquad (11)$$

*By assuming that the annual number of individual waves is $0.5 \times 10^7$, the wave corresponding to N years (fifty-year or one hundred-year, etc.) is given by:*

$$\tilde{H}_{ext(N)} = C\, H_c^{1/d} [15.42 + \ln(N)]^{1/D} \qquad (12)$$

*The foregoing approach raises the following problems:*

*(a) The assumption of independence between the observations $x_1 \ldots, x_n$, involving individual waves, is obviously not satisfied.*

*(b) The use of Equation (1) $F_{X_{ext(n)}}(x) = [F_X(x)]^n$ to obtain the extreme value distribution of a sample of n observations would be strictly correct if the distribution function of the parent law was known ($F_X(x)$). However, since it is not known, it is determined by combining statistical adjustments with mathematical values relying on specific assumptions. The uncertainty in the model may therefore give rise to substantial errors in the extreme parameters estimated.*

*Furthermore, the definition of the extreme parameters given in Section 2.1.1 assumes that one can calculate the mode and dispersion characteristics of the extreme distribution whose density is given by the Equation (8). This equation serves to calculate the dispersion characteristics numerically, but the advantage offered by asymptotic methods is to provide a simple expression of these characteristics, and above all, the fact that they are far less dependent on any modelling errors in $F_X(x)$.*

*These considerations show the drawbacks of modelling the crest-to-trough height from all the individual waves. In brief, the essential criticism of approach A resides in the fact that the high values observed are ignored, because the long-term model of H gives priority to the central values of the distribution.*

*Reference [2.2] gives an example taken from a work by Gumbel which illustrates the dangers incurred by the improprer use of the statistics of extremes. We shall*

It is assumed that the parent distribution law is a logistic form $F(x) = \dfrac{1}{1 + [exp(-\pi x/\sqrt{3})]}$) with a mean $E(X) = 3$ and variance $V(X) = 2.25$. The deviation between the logistic law and the normal law $N(0,1)$ is very small since $sup_x [F(x) - \Phi(x)] \simeq 0.228$.

If the statistician models X by a normal distribution, he does not commit a large error in evaluating the central part of the distribution. It can even be assumed that the *model* adopted has exactly the same mean and the same variance as the logistic distribution $N(3, 2.25)$. If a certain decision is associated with the probability that the maximum $X_{ext(100)}$ on a sample of 100 observations is lower than a given high value, 7.62 for example, it is necessary to calculate $Prob(X_{ext(100)} < 7.62)$. However, according to the normal distribution:

$$Prob(X_{ext(100)} < 7.62) = 0.9048$$

whereas the exact logistic distribution gives:

$$Prob(X_{ext(100)} < 7.62) = 0.6879$$

The modelling error being considerable, is liable to lead to the adoption of a decision that would have been discarded if the exact logistic distribution had been known.

Hence the conclusion [2.2]: "In general, infinitesimal modelling errors may exert a considerable influence on the forecasting of extremes, which is essentially a very delicate matter. It is important to be wary in particular of methods which consist in adjusting the central part of a distribution on the sampling data to infer the results on the unobserved tails of the distribution." (Fig. 2.2).

### Asymptotic methods

These methods consist in selecting the highest values of the sample, with the understanding that the values observed here are no longer individual waves but maximum values recorded over a period of T years. Use could be made, for example, of maximum crest-to-trough heights corresponding to 20 recordings covering a period of $T_0$ years. After making this selection, one can use the *theory developed in Section 2.1.2, together with a method for forecasting the extreme parameters relative to a very broad period of T years (T > $T_0$).*

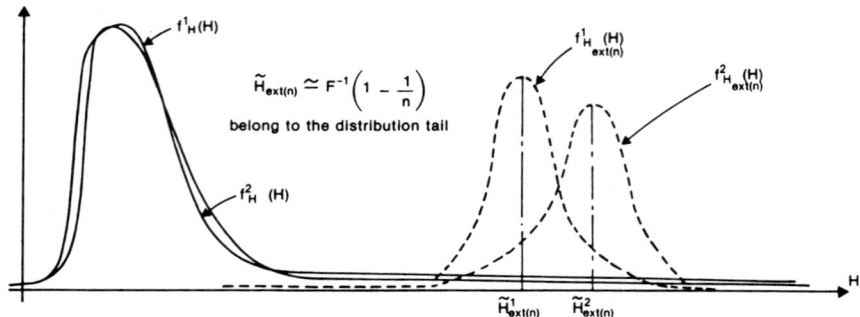

*Fig. 2.2. Possible influence of modelling uncertainty on the extreme value distribution.*

*In particular, if the maximum annual values over a sufficiently long period are available, these values may be considered as observations from the extreme value distribution relative to a one-year period. One can then directly adjust a law of type I, II or III to the realizations observed. This is the modelling approach.*

*Diagram 1, in Section 2.1.4, presents the computation system applicable to the two cases discussed above. The reader should refer to the specialized literature [2.6] for a detailed examination of this methodology. A few remarks are presented here to justify the recommendations in Section 2.1.4.*

**Remark 1:**

*Working with distribution tail (Y = X for X > u) reduces the degree of dependence between the observations $y_1, y_2, \dots y_n$ and hence the influence of this dependence on the results obtained. For the North Sea, for example, it was possible to limit the examination to the recordings corresponding to sea states with the maximum values for the significant heights greater than 5 m, and thus work on a number of observations of about 100.*

*Remark 2:*

The use of the distribution tails as the exclusive basis for the application of the formalism of the theory of extreme values is justified by the need for a better modelling of this tail. The influence of modelling errors is consequently considerably reduced.

A recent study [2.6] showed that a model which attempts to achieve a close adjustment of the tail of the distribution tail gives it substantially different results. The replacement of a Gumbel law by a modified Gumbel law in the study in question led to a 15% increase in the "extreme value predictor". A good model of the higher values observed in the crest-to-trough height H is therefore a necessary condition for obtaining valid asymptotic results.

*Remark 3:*

The semi-analytical methods used for approach A, Section 2.1.3, assumes Gaussian short-term sea states, with zero band width (narrow band spectrum). The asymptotic approach, which is purely statistical, has to discard this assumption, which is generally unrealistic (Fig. 2.3).

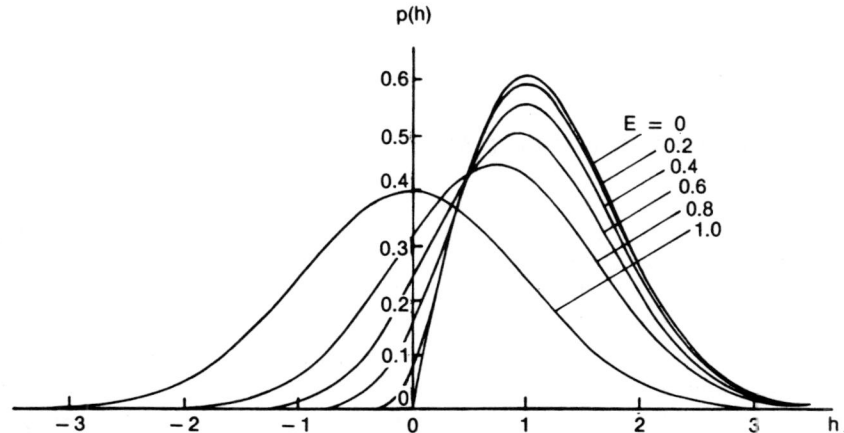

Fig. 2.3. Probability density of peak amplitudes
of the random variable η as a function h of
parameter ε for a given short-term sea
state [2.9].

*Remark 4:*

*The separation between the adjustment of an extreme value distribution over T\* years ($G^{T^\star}$) and the assessment of the relative behavior in T years ($G^T$) highlights the fact that the asymptotic law obtained is associated with the instrumentation time, and that the passage to a duration of T years is a matter of prediction. Approach A makes no distinction between these two steps, whereas the long-term distribution of short-term parameters is actually established from recordings covering fairly short periods. The $H_{1/3}$-T scatter diagram concerning the India station [2.5] is obtained from 2,400 recordings of 12 min each selected from a population of available recordings measured over 13 years from 1952 onwards.*

## 2.1.4 RECOMMENDATIONS

In all cases, except when the modelling approach is feasible, it is recommended to base the statistical analysis of extremes on the distribution tails, i.e. on the highest values effectively observed for the crest-to-trough height.

Due to the conceptual requirement which make the maximum crest-to-trough height H over T years a random variable whose mode and dispersion characteristics are both to be determined (since these characteristics must be as independent as possible of the modelling performed), it is recommended to estimate the behavior of $[F(x)]^n$ for high n by determining the asymptotic laws of extreme values (Fig. 2.4).

Diagram 1 shows a detailed flow-chart of the procedure for determining the extreme wave parameters by asymptotic methods.

## 2.1.5 USE OF AVAILABLE DATA

Approach A requires consideration of all the individual waves, hence of a very large volume of data. The "modelling" approach, which is theoretically the best, is rarely practicable because the data which provide a basis for it (annual maxima over a long period) are actually only available in exceptional cases.

The "distribution tail" has the merit of making statistical treatment possible, even if the available data are not very abundant. In fact, the basic idea is to clarify the information about extreme values obtained in the observation. A small volume of data would therefore allow a prediction of extreme values over a period of T years, while inducing the significance that should be attributed to this prediction. The rarer the information in volume, the more sophisticated the mathematical

Diagram 1
Calculation flow-chart for the use of asymptotic methods

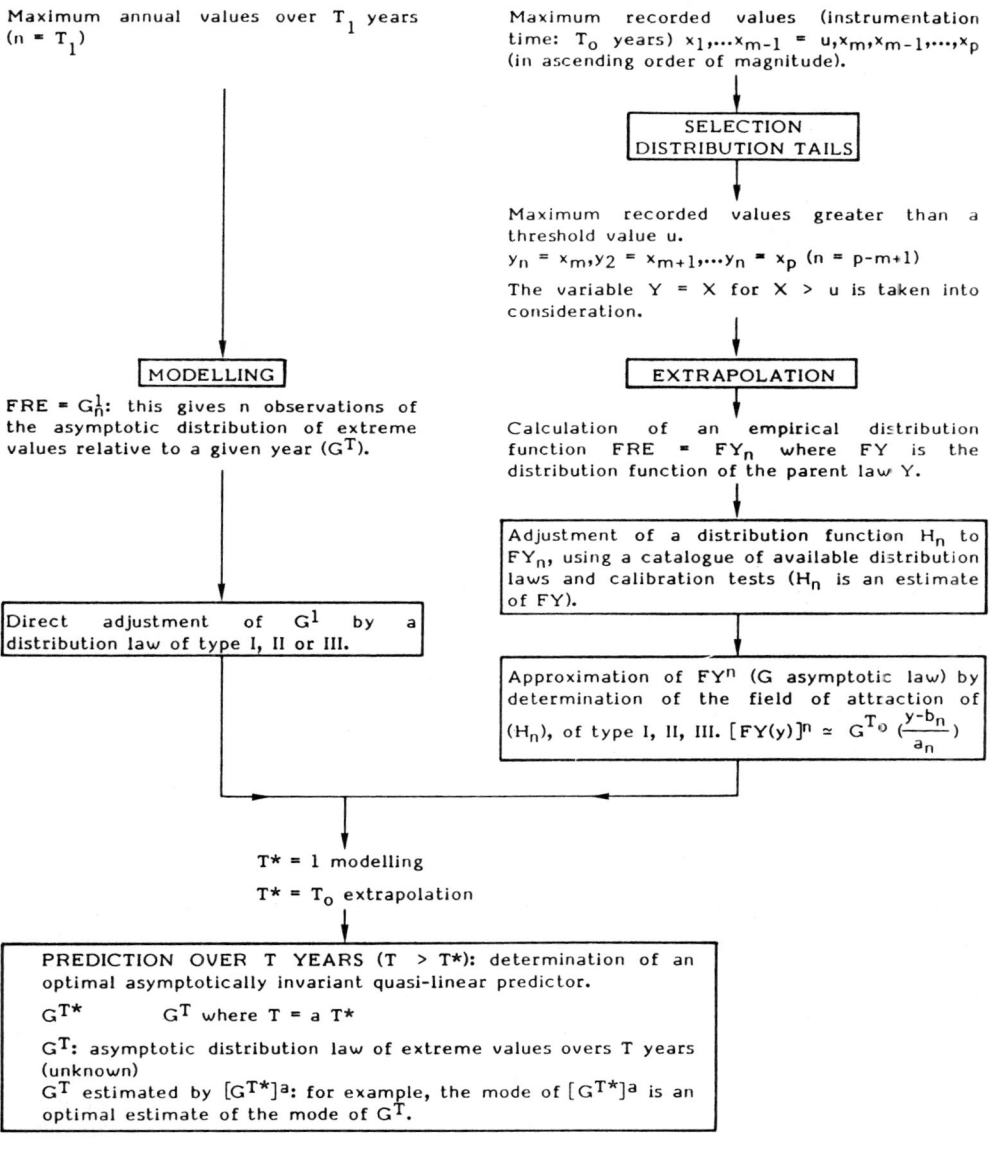

Maximum annual values over $T_1$ years
$(n = T_1)$

Maximum recorded values (instrumentation time: $T_0$ years) $x_1, \ldots x_{m-1} = u, x_m, x_{m-1}, \ldots, x_p$ (in ascending order of magnitude).

SELECTION
DISTRIBUTION TAILS

Maximum recorded values greater than a threshold value u.
$y_n = x_m, y_2 = x_{m+1}, \ldots y_n = x_p$ $(n = p-m+1)$

The variable $Y = X$ for $X > u$ is taken into consideration.

MODELLING

FRE $= G_n^1$: this gives n observations of the asymptotic distribution of extreme values relative to a given year $(G^T)$.

EXTRAPOLATION

Calculation of an empirical distribution function FRE $=$ $FY_n$ where FY is the distribution function of the parent law Y.

Adjustment of a distribution function $H_n$ to $FY_n$, using a catalogue of available distribution laws and calibration tests ($H_n$ is an estimate of FY).

Direct adjustment of $G^1$ by a distribution law of type I, II or III.

Approximation of $FY^n$ (G asymptotic law) by determination of the field of attraction of $(H_n)$, of type I, II, III. $[FY(y)]^n \simeq G^{T_0}\left(\frac{y-b_n}{a_n}\right)$

$T^* = 1$ modelling

$T^* = T_0$ extrapolation

PREDICTION OVER T YEARS (T > T*): determination of an optimal asymptotically invariant quasi-linear predictor.

$G^{T^*}$        $G^T$ where $T = a\ T^*$

$G^T$: asymptotic distribution law of extreme values overs T years (unknown)
$G^T$ estimated by $[G^{T^*}]^a$: for example, the mode of $[G^{T^*}]^a$ is an optimal estimate of the mode of $G^T$.

developments, if significant results are to be inferred.  Hence it is important to identify clearly what can be obtained from the sample and what is provided by the modelling assumptions.

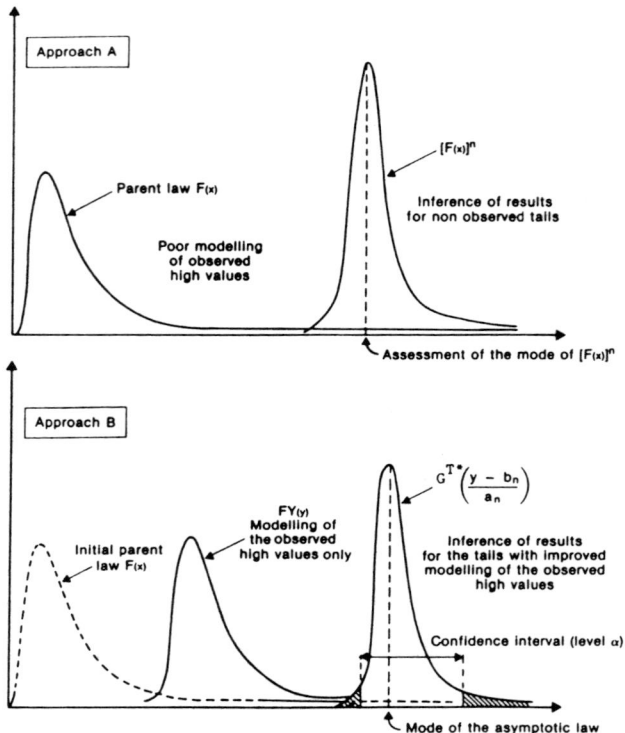

Fig. 2.4. Illustration of differences between the two procedures given in Section 2.1.

## 2.1.6  RANGE  OF PERIODS ASSOCIATED WITH EXTREME PROJECT PARAMETERS

A range of periods is associated with the extreme project parameters defined according to the procedure recommended in Section 2.1.4.  In fact, in a given sea state, a continuous distribution of periods is obtained for a given height, and different meteorological conditions may give rise to sea states with identical extreme wave heights, but with different associated periods.

At the present time, in the North Sea, for a 30 m project wave, the range of associated periods lies between 12 and 18 s.  The determination of the range of periods associated with project design extreme parameters is based on one of the following:

1. The observed data for the variable $H_{max}/(TH_{max})^2$ where $H_{max}$ is the highest wave, counted between 2 passages through zero by increasing value, recorded during a storm, and $TH_{max}$ is the associated period. This method assumes the availability of a large body of storm data always recorded continuously.

2. The joint distribution of wave height (individual)-period for storm cases. Starting with the Gaussian model of the sea state (see Section 2.1, Part III), certain authors [ 2.15 ] have given an analytical expression of the height-period joint probability distribution law for individual waves:

H is the crest-to-trough height,
T is the associated period,
ε is the band width parameter.

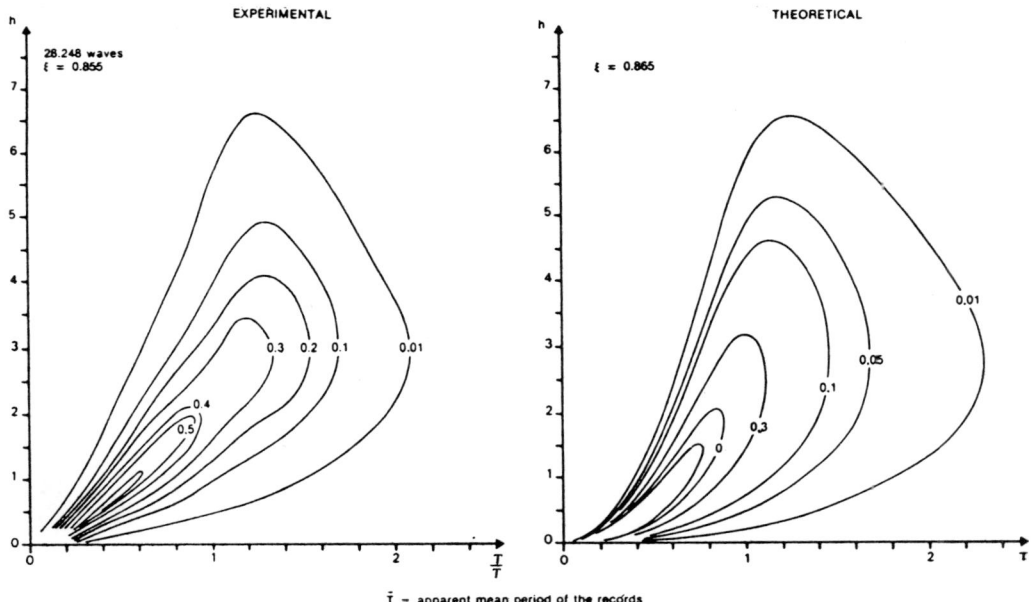

Fig. 2.5. Experimental and theoretical isodensity
curves relative to the joint wave height
(individual) - associated period distribution.

This model is written as:

$$p(h, \tau_1, \varepsilon) =$$

$$\frac{\alpha^3 \ h^2}{4(2\pi)^{\frac{1}{2}} \ \varepsilon(1-\varepsilon^2) \ \bar{\tau}^4 \tau_1^5} \ \exp\left\{-\frac{h^2(\bar{\tau}\tau_1)^4}{8\varepsilon^2} \left[(\bar{\tau}^2\tau_1^2 - \alpha^2)^2 + a^2\alpha^4\right]\right\} \quad (13)$$

where $h = H/\sqrt{m_0}$ and $\tau_1 = \dfrac{T}{\overline{T}\ \overline{\tau}(\varepsilon)} = \dfrac{\tau}{\overline{\tau}(\varepsilon)}$, $\alpha$, a and T are the standard parameters used in modelling the random wave. $\overline{\tau}(\varepsilon)$ is the mathematical likelihood of variable $\tau$, defined by:

$$\overline{\tau}(\varepsilon) = \int_0^\infty \int_0^\infty \tau p(\eta,\tau,\varepsilon)\ d\eta\ d\tau \qquad (h = 2\eta)$$

Figure 2.5 gives the isodensity curves corresponding to Equation (13) and the experimental isodensity curves plotted from 28,240 wave recordings obtained from the recording of 200 storms in the North Sea.

*Model (13) was established by considering a relation of the sinusoidal type between the signal $\xi_1$ and the second derivative of the peak $\xi_3$. In other words, starting with the joint signal density $p(\xi_1, \xi_3)$ given by Cartwright and Longuet-Higgins, a sinusoid with amplitude $\xi_1$ and period T given by $T = 2\pi\sqrt{-\xi_1/\xi_3}$ is associated to every wave amplitude $\xi_1$.*

*This approach is only strictly valid for a zero value of the band width parameter ($\varepsilon = 0$).*

## 2.2 WIND, CURRENT, EARTHQUAKES, TIDES

Each of these forces shall be dealt with by considering the most unfavorable case, in other words, the one leading to the highest forces acting on a given joint.

The action of wind on the structure concerns its atmospheric part, as well as the emerged superstructures (decks, drilling derricks, etc.). Modelling of the wind $V(M,t)$ at a given point and a given instant is specific to this type of action, and involves the mean velocity $\overline{V}(M)$ (over a time interval generally lying between 10 min and 1 hour) as well as $\Delta V(M,t)$ which represents the rapid fluctuation of $V(M,t)$ about $\overline{V}(M)$:

$$V(M,t) = \overline{V}(M) + \Delta V(M,t)$$

The mean velocity $\overline{V}(M)$ at height Z may be expressed as a function of the mean velocity $\overline{V}(MR)$ at a reference point of the height $ZR = 10$ m (33 feet) by a power law such that:

$$\overline{V}(M) = \overline{V}(MR) \, (Z/ZR)^{1/N}$$

where $1/N$ is an exponent between $1/13$ for gusts and $1/8$ for sustained winds in the open sea.

Combined with the theory of extreme values and the methodology presented in Section 2.1, the specific modelling may culminate in a new approach for determining the extreme wind velocities, liable to occur at a given site and at a given height. Unlike the case of extreme values of the crest-to-trough wave height H, no systematic study is available at the present time lying within this framework. The lack of data concerning maximum wind velocity values at sea partly explains this deficiency.

*The BSI standard [4.8] gives project design wind velocity values for use in the North Sea.*

## REFERENCES

2.1    Gumbel, E.J., Statistics of Extremes, Columbia University Press, New York, 1958.

2.2    Deheuvels, P., Théorie classique des extrêmes (cas unidimensionnel), ARAE Extreme Values Seminar, Centre Oceanologique de Bretagne, ARAE Report, June 1983, IFP 31656, Paris, 1983.

2.3    Fisher, R.A. and Tippett, L.H.C., Limiting forms of the frequency distribution of the largest or smallest member of a sample, Proceedings Camb. Phil. Soc., 24, 180, 1928.

2.4    Nordenstrom, N., A method to predict long-term distributions of waves and wave-induced motions and loads on ships and other floating structures, DnV Publication No. 81, April 1973.

2.5    Nordenstrom, N., Methods for predicting long-term distributions of wave loads and probability of failure for ships, Part I Environmental conditions and short-term response, DnV Report No. 71-2-S, March 1972.

2.6    Labeyrie, J., Modélisation des paramètres d'environnement et de comportements d'ouvrages en mer, Journées Statistiques et Sciences de l'Ingénieur, ASU, pp. 20-35, May 1984.

2.7    Labeyrie, J., Axiomatique pour la prédiction de valeurs extrêmes, ARAE Report, Ref. IFP 31611, Paris, 1983.

2.8    Labeyrie, J., Statistiques extrêmes des hauteurs de vagues de tempête, ARAE Report, Ref. IFP 33146, Paris, 1985.

2.9    Ochi, M.K., On prediction of extreme values, Journal of Ship Research, Vol. 17, 1973.

2.10   Tiago de Oliveira, J. and Gomes, I., Two test statistics for choice of univariate extreme models, NATO Adv. Stud., International Seminar, Vimeiro, 1983.

2.11   Pickands, J., Statistical inference using extreme order statistics, Arm. Stat., Vol. 3, 1975.

2.12   David, M.A., Order Statistics, J. Wiley and Sons, New York, 1970.

2.13   Smith, R.L., Uniform rates of convergence in extreme value theory, Adv. Appl. Prob., Vol. 14, 1982.

2.14 Aagaard, P.M. and Petrauskas, C., Extrapolation of historical storm data for estimating design wave heights, OTC 1970, Paper No. 1190.

# Determination of Load Cases
# Requiring Verification

The analysis of the static strength of welded tubular joints must be conducted for all of the conditions likely to be encountered during the life of the structure. The load cases differ essentially in the severity of the forces applied to the structure by the environment (wind, wave, etc.).

## 3.1  NORMAL CONDITIONS

In normal operating conditions, also called "operational" conditions, the environmental forces have no influence on the operation of the platform.  The types of load to be accounted for are:

(a) Dead weight of the structure.

(b) Non-permanent loads, considering both their maximum effects and their minimum effects.

(c) Forces due to the environment: hydrodynamic forces, hydrostatic thrust, wind effect, etc.

## 3.2  EXTREME CONDITIONS

These are related to the extreme values of the climatic and oceanographic parameters determined in Section 2.  In these conditions, certain operations are interrupted (drilling activities, movement of the derrick, use of the cranes) while others continue (mud treatment, various handling operations).

The types of load to be accounted for are:

(a) Dead weight of the structure.

(b) Non-permanent loads compatible with extreme conditions, considering both their maximum effects and their minimum effects.

(c) Extreme forces due to the environment and the hydrostatic thrust.

## 3.3 TEMPORARY CONDITIONS

Temporary conditions are those prevailing during the construction, towing, installation and any other operation likely to take place during the life of the structure.

The type of load to be accounted for are:

(a) Dead weight of the structure.
(b) Maximum temporary loads specific to the phase considered (towing, installation, etc.).
(c) Forces due to the environment (wind, wave, hydrostatic thrust) corresponding to the phase considered.

*During the phases preceding their commissioning, the platforms and welded tubular joints are subject to stresses that are sometimes greater than those exerted after they enter service. In this case, the specific temporary conditions are used for designing certain structural details.*

*Examples:*

*1. During the launching phase; certain joints are more deeply submerged than in their final position, and are therefore subject to greater forces.*

*2. On site installation phase; if the structure is launched over the stern of the barge, considerable forces are generated at certain points, because the structure overhangs the barge deck.*

**Remarks:**

The different types of loading are combined in accordance with the probability of their simultaneous occurrence. With respect to combinations of environmental loads, the regulations specify the minimum combinations to be taken into account for each of the foregoing conditions. Standard practice consists in superimposing the extreme value of the crest-to-trough height H with the extreme value of the wind speed. It is improbable that these two extreme values will occur simultaneously. The wind speed corresponding to the extreme value of H is slightly lower than extreme value. In these conditions, since the wind is less of a design factor than the wave, the superimposition of extreme values ensures safety without being too conservative.

CHAPTER **4**

# Ultimate Static Strength Formulas for Welded Tubular Joints

## 4.1 PROCEDURE FOR ESTABLISHING ULTIMATE STATIC STRENGTH FORMULAS FOR WELDED TUBULAR JOINTS SUBJECT TO SIMPLE LOADS

Static resistance formulas based on the analysis of test results associate the nominal ultimate loading with the geometric and mechanical characteristics of the assembly. The development of these formulas implies:

(a) The definition of a failure criterion for each type of joint geometry and each type of load.

(b) The determination of an analytical model for the ultimate static strength which, depending on each particular case, assumes a more or less empirical character. To formulate the final analytical model, in addition to obvious lessons drawn from experience, use is made of simplified physical models, discretisation methods, and specific theories (strength of materials), which are presumed to reflect reality and which may be quite different in nature.

(c) Statisical treatment of the tests results, namely:

. Determination of a reference population including a uniform group of joints corresponding to the failure criterion defined in (a).

. The use of statistical methods (multiple regression) which serve to assess the average behavior of a given joint, subjected to a given load, as well as the random residual deviation likely to occur from this average behavior. The statistical treatment consists of adjusting the coefficients of the analytical model determined in (b).

(d) The insertion of results obtained in (c) into a semi-probabilistic approach to safety and design, which must take account as much as possible of all the uncertainties other than those corresponding to the statistical treatment of test results.

Several methodologies for the semi-probabilistic approach to safety are available, including:

(a) The method recommended by ISO[1] and ECCS[2] (Fig. 4.1).

(b) The method recommended by ASCE[3] called LRFD (Load and Resistance Factor Design), and which is based on the determination of the safety index $\beta$ (Fig. 4.2).

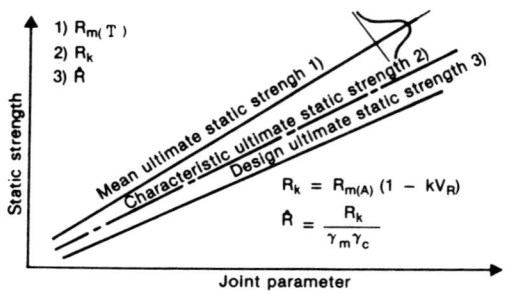
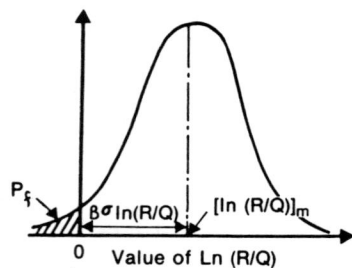

Fig. 4.1. Approach recommended by ISO and ECCS.

Fig. 4.2. Safety index $\beta$ in the LRFD approach.

(1) ISO   : International Standards Organisation
(2) ECCS : European Convention for Constructional Steelwork
(3) ASCE : American Society of Civil Engineers

A distinction is made between the following:

(a) The mean ultimate static strength (tests) $R_{m(T)}$, resulting from the statistical treatment of test results. The dimensions of the geometric and mechanical parameters are the actual (measured) dimensions, at least as concerns $T$, $\sigma_y$ and $\sigma_y/\sigma_u$.

(b) The "mean" ultimate static strength $R_{m(A)}$ obtained from the foregoing but considering the random character of the geometric and mechanical parameters. This is justified because the calculation is conducted at the design stage, i.e. on the basis of nominal values.

(c) The nominal ultimate static strength $R_n$ provided by the static resistance formula (test results), by using the nominal values of the parameters concerning the joint.

(d) The characteristic ultimate static strength $R_k$, associated with a given probability (95%, 97.5%) and involving uncertainties connected with statistical treatment, as well as those deriving from the random character of the geometric and mechanical parameters. This characteristic strength is a function of the mean $\bar{R}_{m(A)}$ and of the coefficient of variation $V_R$ of the "mean" ultimate static strength $R_{m(A)}$.

(e) The design ultimate static strength, R, derived from the foregoing after using safety factors which account for other sources of uncertainty.

Verification formula for the static strength of the joint is then written:

(a) For the method recommended by ISO and ECCS:

$$\Sigma(\gamma_{sk}Q_k) \quad \hat{R} = \frac{R_k}{\gamma_m\gamma_c}$$

(b) For the LRFD method:

$$\Sigma(\gamma_{sm}Q_m) < \hat{R} = \Phi R_n$$

where

$$\Phi = e^{-\alpha\beta V_R} \cdot \frac{\bar{R}_{m(A)}}{R_n}$$

$\gamma_m$   is a partial safety coefficient for the material,

$\gamma_c$   is a partial safety coefficient which takes account of the type of structure and its behavior, as well as the degree to which it is critical that a certain limit state is reached,

$\alpha$   is a coefficient, equal to 0.55, specific to the LRFD method,

$\beta$   is "calibrated" to obtain the same level of safety as that of existing codes. For welded tubular joints, the mean value of $\beta$ is normally taken as 4.5,

$Q_m, Q_k$   are the mean and characteristic values of the loads respectively,

$\gamma_{sm}, \gamma_{sk}$   represent the values of the partial safety coefficients for the mean and characteristic loads respectively.

For the determination of R, the use of either of the two approaches requires the following:

(a) The assessment of the characteristics (mean, coefficient of variation) of the basic variables: T, D, $\sigma_y$ and $\sigma_y/\sigma_u$.

(b) The determination, by a probability calculation, of the characteristics (mean, coefficient of variation) of the mean ultimate static strength: $R_{m(A)}$, or $\overline{R}_{m(A)}$ and $V_R$.

(c) The choice of a value to be given to the coefficients $\gamma_m$ and $\gamma_c$.

Statistical treatment of test results serves to recommend the use of certain formulas for the ultimate static strength $R_{m(T)}$. On the other hand, the choice of the values of $\gamma_m$ and $\gamma_c$ depends on specific decisions related to the desired safety level.

**A.** *Several existing regulations rely on the concept of punching shear (Fig. 4.3). In this approach, one verifies that the value of the punching shear corresponding to the ultimate load is lower than the admissible punching shear, the latter being determined from test results.*

$$v_p \leqq V_p \quad where \quad v_p = \frac{t}{T} \frac{f_a \sin \Theta}{k_a} + \frac{f_b}{k_b}$$

$$and \quad V_p = Q_q \, Q_p \, Q_f \frac{F_y}{0.9 \, \gamma^{0.7}}$$

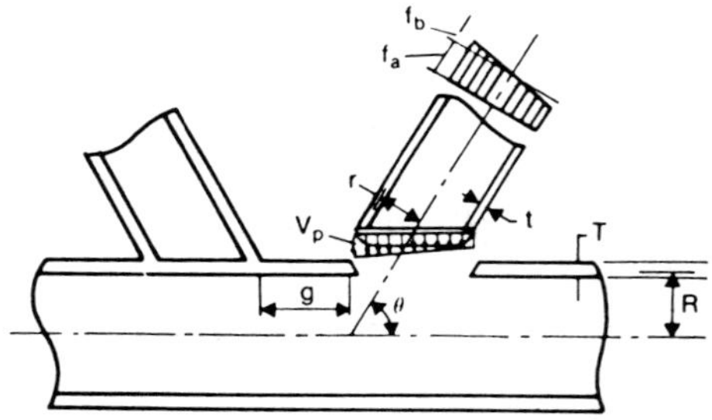

*Fig. 4.3. Punching shear method.*

(a) *$k_a$ and $k_b$ are length and cross-section factors related to the intersection of the tubes making up the assembly.*

(b) *$F_y$ is the yield strength of the chord steel. It is taken as 2/3 $\sigma_u$ if this value is less than $\sigma_y$.*

(c) *$Q_p$ is the plastic reserve factor taking account of the favorable interaction in the event that two or*

more types of load are applied simultaneously to
the joint.

(d) $Q_q$ is a factor associated with the geometry and
type of loading.

The procedure proposed here in the recommendations
appears more valid insofar as it requires fewer
assumptions on the stress distribution, which is far more
complex than the distribution adopted in the punching
shear approach.

**B.** Reference [ 4.79 ] gives a presentation of 864 static
strength tests taken from the literature. These are
mostly Japanese tests (75%). Even though the general size
of the joints is smaller than that of real structural
joints, many of the values of the geometrical parameters
β and γ are representative. Table 4.1 identifies these
joints by type of geometry and type of load.

Table 4.1.
Identification of joints by type of geometry
and type of load [ 4.79 ].

| | T | Y | K | N | X (θ=90°C) | X | KT | |
|---|---|---|---|---|---|---|---|---|
| Tension | 41 | 3 | – | – | 51 | 4 | – | 99 |
| Compression | 92 | 9 | – | – | 88 | – | – | 189 |
| Axial | – | – | 408 | 97 | – | | – | 12 | 517 |
| In-plane bending | 39 | – | 2 | – | – | – | – | 41 |
| Out-of-plane bending | 12 | 2 | 4 | – | – | – | – | 18 |
| Total | 184 | 14 | 414 | 97 | 139 | 4 | 12 | 864 |

The mode of load application of K and N joints is
varied. As a rule, a brace was loaded in tension or
compression, and the load on the other brace corresponded
to a reaction. If the load was applied in tension, the
ultimate load on the other brace (compressed) was often
calculated from the ultimate tensile load measured,
assuming a certain load transmission mode. In some cases,
both braces were loaded directly. Statistical treatment
does not account for this heterogeneity, which may be
considered as partly responsible for the random residue.

**C.** *Formulas for mean ultimate static strength of welded tubular joints under simple loads (axial loading of K and N joints).*

*The formulas given in [4.66, 4.67, 4.75 and 4.76] are the only ones based on the multiple regression method, and correspond to the lowest scatter of the test results. They are presented in Table 4.2 with their range of validity $D_V$. The functions $f_1$ to $f_7$ are related to the following general equation:*

$$R_{m(T)} = f_1(\beta).f_2(\gamma).f_3(\Theta).f_4(g/T).f_5(r).f_6(\sigma_y/\sigma_u).f_7(\alpha)\sigma_y T^2$$

*where r is the ratio of the nominal stress in the chord to the yield strength $\sigma_y$.*

*The minimum and maximum values of the different parameters are not sufficient to determine the validity range, which is only part of the multidimensional space effectively containing the test results (Fig. 4.4).*

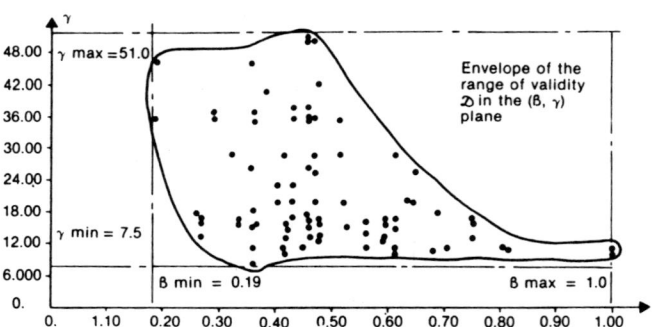

*Fig. 4.4. Plot of β versus γ for 418 K and N tubular joints.*

Table 4.2.
Formulas for the mean ultimate static strength (tests) $R_{m(T)}$

| | X joints | $D_v^{(1)}$ | T and Y joints | $D_v^{(1)}$ | K and N joints | $D_v^{(1)}$ |
|---|---|---|---|---|---|---|
| $f_1(\beta)$ | $\dfrac{7.46}{1-0.812\beta}$ | 0.12-1.0 | $4.76(1+4.93\,\beta^2)$ | 0.19-1.0 | $2.65(1+4.43\beta)$ | 0.19-1.0 |
| $f_2(\gamma)$ | $(2\gamma)^{-0.05}$ | 8-51 | $(2\gamma)^{0.241}$ | 8.5-47 | $(2\gamma)^{0.181}$ | 7.5-51 |
| $f_3(\theta)$ | $1/\sin\theta$ | 90° | $1/\sin\theta$ | 45°-90° | $\dfrac{(1-0.319\cos^2\theta)}{\sin\theta}$ | 30°-90° |
| $f_4(g/T)$ | 1.0 | | 1.0 | | $1+\dfrac{0.005(2\gamma)^{1.49}}{1+\exp(0.355g/T-0.7)}$ | -15 to +33 |
| $f_5(r)$ | $1.22-0.5\lvert r\rvert$<br>$1.0$ if $r\geq-0.44$ | | 1.0 | | $1+0.302r-0.283r^2$ | |
| $f_6(\sigma_y/\sigma_u)$ | $(\sigma_y/\sigma_u)^{-0.173}$ | 0.65-0.94 | 1.0 | 0.55-0.87 | $(\sigma_y/\sigma_u)^{-0.739}$ | 0.56-0.94 |
| $f_7(\alpha)$ | 1.0 | | $(\alpha/2)^{-0.461}$ | 6-10 | 1.0 | |
| Size | 76 | | 74 | | 418 | |
| Mean value of ratio $P_{ult(es.)}/R_{m(E)}$ | 1.005 | | 1.0053 | | 1.0024 | |
| Standard deviation of the above ratio | 0.109 | | 0.103 | | 0.108 | |

(1) The limiting values given correspond to the limiting values of the parameters $\beta$, $\gamma$, $\theta$, $g/T$, $r$, $\sigma_y/\sigma_u$ and $\alpha$.

*In principle, static strength formulas cannot be applied (from the statistical standpoint) outside the validity range thus defined. The extension of their application beyond this range is only justified by a hypothesis concerning the physical behavior of the joints in the extension range.*

*Figure 4.5 provides an example of the way in which $f_1(\beta)$ is determined from test results (T and Y joints):*

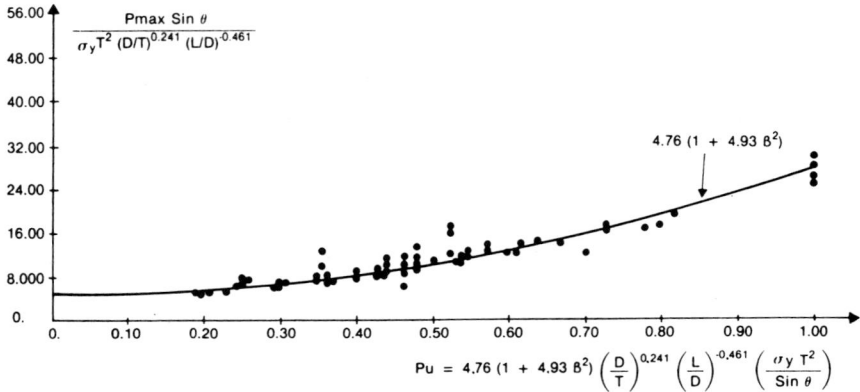

$$Pu = 4.76 \left(1 + 4.93 \, \beta^2\right) \left(\frac{D}{T}\right)^{0.241} \left(\frac{L}{D}\right)^{-0.461} \left(\frac{\sigma_y \, T^2}{\text{Sin } \theta}\right)$$

*Fig. 4.5. Formula for ultimate static strength (tests) for T and Y joints in compression (see Table 4.2).*

*D. The use of a semi-probabilistic approach requires the knowledge of the discrepancies that may exist between the nominal values and the actual values after fabrication.*

*Table 4.3 gives the evaluations of these deviations for the variables D, T, $\sigma_y$ and $\sigma_y / \sigma_u$ [4.67, 4.76] .*

*Table 4.3.*
*Means and coefficient of variation (COV)*
*of basic variables*

| Nominal value | Hot finished tubes | | Cold finished tubes | |
|---|---|---|---|---|
| | Mean | Coefficient of variation | Mean | Coefficient of variation |
| $D_n$ | $1.0\ D_n$ | $0.005$[1] | $1.0\ D_n$ | $0.004$[1] |
| $T_n$ | $1.0\ T_n.$ | $0.05$ | $1.0\ T_n$ | $0.05$ |
| $\sigma_y$[2] | $1.18\ \sigma_y$ | $0.075$ | $2.35(DT)^{0.121}\sigma_y$ | $0.11$ |
| $\sigma_y/\sigma_u$ | $0.66$ | $0.13$ | $0.92(D/T)^{0.039}$ | $0.035$ |

(1) The variation of the outside diameter of the chord, D, has been neglected in the establishment of the design strength formulas.

(2) $\sigma_y$ is the minimum guaranteed yield strength.

## 4.2  DESIGN ULTIMATE STATIC STRENGTH FORMULAS
## FOR WELDED TUBULAR JOINTS

It is recommended to use the ISO and ECCS approach, which is also taken into account in the draft Common Unified Rules for Steel Structures, EUROCODE 3. The formulas given in Sections 4.2.1 and 4.2.2 concern hot formed tubes.

### 4.2.1  AXIAL LOADINGS

It is recommended to use the following formulas (Fig. 4.6) in conjunction with the nominal values of the parameters $\beta$, $\gamma$, $\Theta$, $\alpha$, $g$, $r$, $T$. For T, Y and X joints, these formulas were developed from tests on tubular joints under compressive loads.

| Type of joint | f(β) | f(γ) | f(θ) | f(α) | f(g) | f(r) | $\frac{1_0}{Sin\theta}$ | $\sigma_{yk}$ | f(T) |
|---|---|---|---|---|---|---|---|---|---|
| T.Y 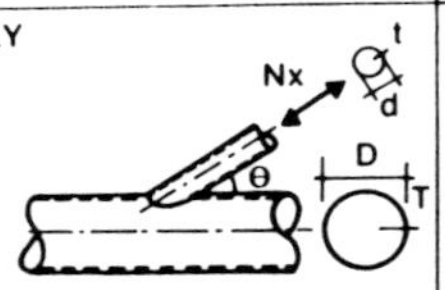 | $\dot{N}_x = (3.96 + 19.53\beta^2)$ | $\gamma^{0.241}$ | 1.0 | $\left(\frac{\alpha}{2}\right)^{-0.461}$ | 1.0 | $f(r)$ | $\frac{1}{Sin\theta}$ | $\sigma_{yk}$ | $T^2$ |
| X 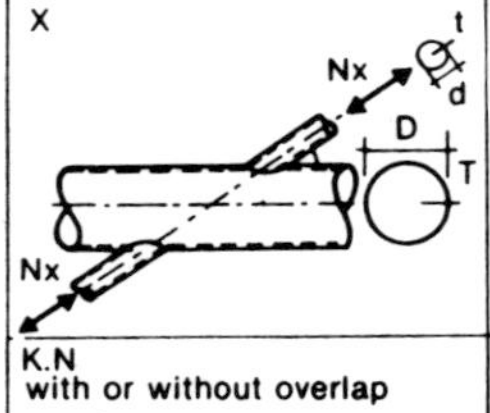 | $\dot{N}_x = \dfrac{5.28}{1-0.812\beta}$ | $\gamma^{-0.05}$ | 1.0 | 1.0 | 1.0 | $f(r)$ | $\frac{1}{Sin\theta}$ | $\sigma_{yk}$ | $T^2$ |
| K.N with or without overlap 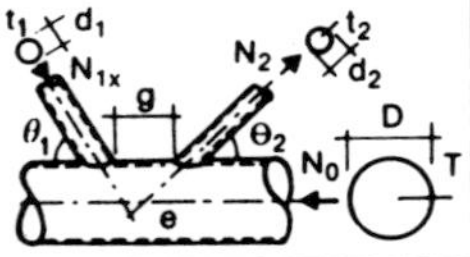 | $\dot{N}_{1x} = (2.61 + 11.56\beta)$ | $\gamma^{0.181}$ | $1 - 0.319 < 0.5^2\theta)$ | 1.0 | $f(g, \gamma)$ | $f(r)$ | $\frac{1}{Sin\theta}$ | $\sigma_{yk}$ | $T^2$ |

with  $f(r) = 1.22 - 0.5\,(r)\;(1.0\;\text{If}\;r \geqslant -0.44)$  Joints X. T and Y

$f(r) = 1 + 0.302\,r - 0.283\,r^2$     Joints K and N

$f(g, \gamma) = 1 + \dfrac{0.00499\,(2\gamma)^{1.49}}{1 + \exp[0.355\,g/T - 0.733]}$

Fig. 4.6. Design ultimate static strength formulas
for axial loads (any geometry).

The formulas recommended were established from mean strength formulas (see Section 4.1) on the basis of the following assumptions:

$$R_k = \overline{R}_{m(A)}(1 - 1.64V_R)$$

where $\overline{R}_{m(A)}$ has a Gaussian distribution and the coefficient 1.64 corresponds to a level of probability of 95%.

$$\gamma_m \gamma_C = 1.25$$

$$\sigma_y = (1-2V_{\sigma_{ym}}) \overline{\sigma}_{ym}$$

In fact, the mean strength formulas (tests) were developed from measured $\sigma_y$ values whereas the design strength formulas employ the guaranteed minimal yield strength.

The value of 1.25 for the product of the partial coefficients $\gamma_m$ $\gamma_C$, which is relatively high, takes account of uncertainties inherent in the method employed. In fact:

(a) It is difficult to establish a connection between tests on small scale models and the actual behavior of the joints of the structure.

(b) Statistical analyses sometimes include very different behaviors in the same treatment, especially concerning ductility. For example, the treatment of K joints draws no distinction between the two main failure modes as shown in Fig. 4.7.

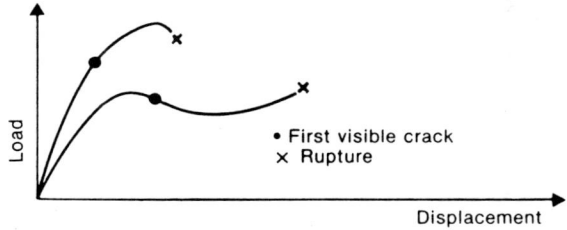

Fig. 4.7. Illustration of two types of load-
         displacement curve observed for K joints.

*The coefficients of variation of the "mean" ultimate static strength $R_{m(A)}$ are:*

*(a) For X joints (compression) $V_R = 0.169$.*

*(b) For T and Y joints (compression) $V_R = 0.155$.*

*(c) K and N joints (axial) $V_R = 0.197$.*

*The coefficient of variation $V_R$ of K and N joints depends on the geometry of the joint. The value 0.197 corresponds to the maximum value calculated in the validity domain $(-20 \leqq g/T \leqq 50$ and $20 \leqq D/T \leqq 100)$.*

*Reference [4.78] gives simplified design ultimate strength formulas. The formulas differ from those recommended here for the following reasons:*

*(a) The elimination of parameters having little influence.*

*(b) A concern for uniformity between the different formulas.*

*(c) A choice of a value of 1.1 for the product $\gamma_m \gamma_c$.*

### Remark:

*The mean formulas $R_{m(T)}$ in Table 4.2, see Section 4.1, are periodically adjusted. The most recent adjustments concerning the formulas adopted in this edition of the guide can be found in [4.75, 4.76, 4.77]. These studies are too recent for a value judgement of their relevance. In fact, they are merely small modifications, while the same theoretical approach as the one adopted here is maintained.*

## Joints under tensile loads

For tensile loaded X, T and Y joints, it is recommended to use formulas for joints with the same geometry under compressive loads.

*The statistical analysis of test results shows that the ultimate static strength of X, T and Y joints under tensile load is greater than that of loads of identical geometry under compression. However, the wide scatter of the results prevents adjustment of formulas as reliable as those presented in Table 4.2. Tensile loaded T and Y joints display greater static strengths than those of X joints also under tensile load and having the same values of the geometrical parameters (D/T, etc.).*

## 4.2.2 IN-PLANE AND OUT-OF-PLANE BENDING LOADS

The following formulas (Table 4.4) are recommended:

Table 4.4.

Design ultimate static strength formulas for
tubular joints under bending loads

| Type of stress | Formulae |
|---|---|
| $M_y$ | $\hat{M}_y = 4.27 \, \beta \, d \, \gamma^{0.5} \, \dfrac{f(r)}{\sin \theta} \, \sigma_y \, T^2$ <br><br> Valid for T, Y and X joints |
| $M_z$ | $\hat{M}_z = 2.82 \, \dfrac{d}{1 - 0.812 \, \beta^2} \, \dfrac{f(r)}{\sin \theta} \, \sigma_y \, T^2$ <br><br> Valid for T, Y, X, K, N joints |
| $f(r) = 1.22 - 0.5|r| \quad (1.0 \text{ if } r \geqslant -0.44$ | |
| Same validity range as for joints under axial loads | |

*In-plane and out-of-plane bend tests concerned T joints almost exclusively.*

*The design ultimate static strength formulas for tubular joints under bending loads are taken from [4.78]. For the methodology of development of these formulas reference should be made to [4.78], which is different from that discussed in Section 4.1. In fact, the small number of test results did not allow the adjustment of mean ultimate strength formulas by multiple regression.*

## 4.2.3 COLD FORMED TUBULAR JOINTS

The formulas given in Sections 4.2.1 and 4.2.2 concern hot formed tubular joints. Similar formulas can be developed for cold formed joints. It suffices to recalculate the "mean" ultimate static strength properties ($\bar{R}_{m(A)}$ and $V_R$) taking for $\sigma_y$ and $\sigma_y/\sigma_u$ the characteristics values of cold formed tubular sections (Table 4.4). The design strengths obtained are greater than those corresponding to hot formed tubes. Consequently, it is recommended to use formulas for hot formed tubes, whatever the tube forming method. For cold formed tubes, this recommendation is conservative.

*The coefficients of variation of mean ultimate static strength are:*

*(a) For X joints (compression) $V_R = 0.186$.*

*(b) For T and Y joints (compression) $V_R = 0.174$.*

*(c) For K and N joints (axial) $V_R = 0.200$.*

*The coefficient of variation of K joints depends on the joint geometry. The value 0.200 corresponds to the maximum value calculated in the validity range: ($-20 \leq g/T \leq 50$ and $20 \leq D/T \leq 100$).*

## 4.3  ULTIMATE STATIC STRENGTH OF WELDED TUBULAR JOINTS UNDER COMPLEX LOADS

For complex loads, it is recommended to use the following linear interaction rule:

$$\frac{N_x}{\hat{N}_x} + \frac{M_y}{\hat{M}_y} + \frac{M_z}{\hat{M}_z} \leq 1$$

where

$N_x$ = design (load factored) axial load exerted on the brace,

$M_y$ = in-plane bending moment exerted on the brace,

$M_z$ = out-of-plane bending moment exerted on the brace,

$\hat{N}_x$ = design axial ultimate strength determined in Section 4.2,

$\hat{M}_y$ = ultimate in-plane bending moment strength determined in Section 4.2,

$\hat{M}_z$ = ultimate out-of-plane bending moment strength determined in Section 4.2.


*Very few static strength tests have been conducted on welded tubular joints under complex loads. Only a few tests have been performed [4.60], in which in-plane bending and axial loads are superimposed for T joints. For three values of the parameter β, the results appear to correspond to the linear interaction rule (Fig. 4.8).*

*Some authors [4.14] have suggested taking account of a favorable effect deriving from the interaction between several types of load. This effect, which is based on the assumption of uniform global plastification at the joint, gave rise to a correction factor $Q_p$ introduced in 1978 in the API regulation [4.13]. Inasmuch that it is not evident that the original assumptions of this approach had been vindicated irrespective of the type of assembly considered, this is not recommended (Fig. 4.9).*

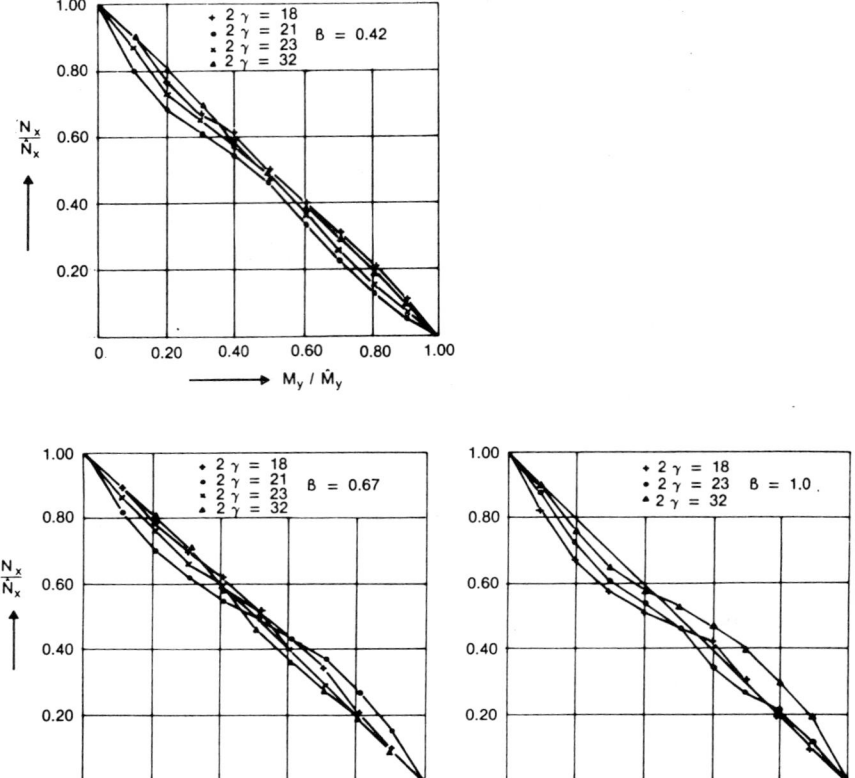

Fig. 4.8. Static test results for T joints
under complex loads: axial load plus
in-plane bending moment [4.60].

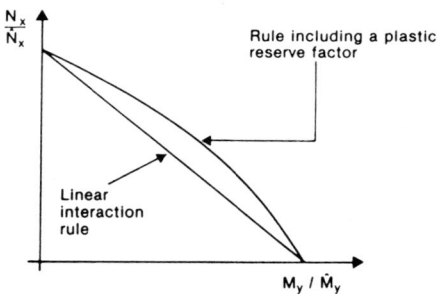

Fig. 4.9. Linear interaction rule and rule
including a plastic reserve factor.

## REFERENCES

**Formula development methodology**

4.1     Strating, J., The interpretation of test results for a Level-1 Code, IIW Doc. XV-452-80/XII-974-80, Stevin Report 6-80-8, TNO, Stevin Laboratory, Department of Civil Engineering, Delft University of Technology, 1980.

4.2     Wardenier, J., Hollow Section Joints, Chap.III, Delft University Press, 1982.

4.3     Draper, N.R. and Smith, H., Applied Regression Analysis, John Wiley and Sons, New York, London and Sydney, 1966.

4.4     Ravindra, M.K. and Galambos, I.V., Load and resistance factor design for steel, Journal of the Structural Division, Proceedings of the American Society of Civil Engineers, Vol.104, No.ST9, September 1978.

4.5     ECCS Recommendations for Steel Construction, ECCS-EG-76-1E, European Convention for Constructional Steelwork, 1978.

(References should also be made to the articles by Y. Kurobane, which deal with both the methodology for the establishment of formulas, and the statistical analysis of the test results).


**Codes and regulations**

4.6     Recommended Practice for Planning, Designing and Constructing Fixed Offshore Platforms, American Petroleum Institute, API RP 2A, 12th Edition, January, 1981.

4.7     Structural Welding Code, American Welding Society, ANSI/AWS D1.1-81, 1981.

4.8     Code of Practice for Fixed Offshore Structures, British Standards Institution, BS 6235:1982.

4.9     Offshore Installations, Guidance on Design and Construction, Department of Energy, HMSO, London, July 1977.

4.10    Rules for the Design, Construction and Inspection of Fixed Offshore Structures, Appendix C, Steel Structures, Det Norske Veritas, 1977.

4.11   Construction métallique, Assemblages soudés de profils creux circulaires avec découpes d'intersection, Conception et vérification des assemblages, AFNOR, NF P 22-250, June 1978.

4.12   Regulations for the Structural Design of Fixed Structures on the Norwegian Continental Shelf, Norwegian Petroleum Directorate, 1977.

4.13   Marshall, P.W., A review of American criteria for tubular structures and proposed revisions, IIW Doc. XI-405-77.

4.14   Lee, M.S., Cheng, A.P., Sun, C.T. and Lair, R.Y., Plastic consideration on punching shear strength of tubular joints, OTC 1976, Paper No.2641.


**Reports on test results**

4.15   Bouwkamp, J.G., Behavior of tubular truss joints under static loads, Phase I, University of California, July 1965.

4.16   Bouwkamp, J.G., Research on tubular connections in structural work, Welding Research Council Bulletin 71, New York, August 1961.

4.17   Bouwkamp, J.G., Behavior of tubular truss joints under static loads, Phgase II, Report No.67-33, Structural Engineering Laboratory, University of California, December 1967.

4.18   Brown and Root, An investigation of welded tubular joints loaded by axial and moment loads, Job No.ER-0169, Offshore Structures Department, Houston, February 1976.

4.19   Gibstein, M.B., Static strength of tubular joints, Det Norske Veritas Report No.73-86-C, May 1973.

4.20   Gibstein, M.B., The static strength of T-joints subjected to in-plane bending, Det Norske Veritas Report No.76-137, April 1976.

4.21   Grigory, S.C., Experimental determination of the ultimate strength of tubular joints, Project No.03-3054, Southwest Research Institute, San Antonio, Houston, September 1971.

4.22   Hlavacek, V., Strength of welded tubular joints in lattice girders, Costruzioni Metalliche, No.6, 1970.

4.23   Kanatani, H., Experimental study of welded tubular connections, Memoirs of the Faculty of Engineering, Kobe University, No.12, 1966.

4.24    Kurobane, Y., Welded truss joints of tubular structural members, Memoirs of the Faculty of Engineering, Kumamoto University, Vol.XII, No.1, December 1964.

4.25    Kurobane, Y. and Makino, Y., Local stresses in tubular truss joints, Research Report, No.14, Kyushu Branch of AIJ, February 1965 (in Japanese).

4.26    Kurobane, Y., Makino, T. and Mitsui, U., Unpublished test results obtained by the authors in Kumamoto University from 1976 until 1979.

4.27    Kurobane, Y., Makino, Y., Honda, T. and Mitsui, Y., Additional tests on tubular K-joints with CHS members under static loads, IIW Doc.XV-460-80, June 1980.

4.28    Mitsui, Y., Experimental study on local stress and strength of tubular joints in steel, Doctoral Dissertation, Osaka University, December 1973 (in Japanese).

4.29    Development of computer program for fatigue design of tubular joints for floating offshore structures, Mitsui Engineering and Shipbuilding Co. Ltd., Unpublished Report, 1978 (in Japanese).

4.30    Nakajima, T. et al, Experimental study on the strength of thin wall welded tubular joints, International Institute of Welding, Doc.No.XV-312-71, Tokyo, 1971.

4.31    Nishida, Y., Sakamoto, S., Ohtake, F. and Minoshima, N., Method for reinforcing tubular truss joints (Investigation into optimum area of chord wall with partially increased thickness), Summary Papers, Annual AIJ Conference, October 1978 (in Japanese).

4.32    Novikov, V.I., Kovtunenko, V.A., Paton, E.O. and Shumitskii, O.I., Direct joining of tubular section components, Automatic Welding, Vol.9, pp. 61-68, 1959.

4.33    Ohtake, F., Sakamoto, S., Tanaka, T., Kai, T., Nakazato, T. and Takizama, T., Static and fatigue strength of high tensile strength steel tubular joints for offshore structures, OTC Proceedings, Paper No.3254, May 1978.

4.34    Popov, V.S., Research into the strength of the joints between the lattice members and chords in tubular welded structures, Aut. Svarkea, No.3, pp. 30-31, 1972.

4.35    Reber Jr., J.B., Ultimate strength design of tubular joints, OTC, paper No.1864, May 1972.

4.36  Sakamoto, S., private communication, Central Research Laboratory, Sumitomo Metal Industries Co. Ltd., 1976.

4.37  Sammet, H., Die Festigkeit Knotenblechloser Rohrverbindungen im Stahlbau, Schweisstechnik, Zeitschrift für alle Gebiete der Schweiss-, Schneid- und Lottechnik, 13, 1963.

4.38  Tebbett, I.E., Beckett, C.D. and Billington, C.J., The punching shear strength of tubular joints reinforced with a grouted pile, Offshore Technology Conference, OTC, Paper No.3463, Texas, 1979.

4.39  Togo, T., Experimental study on mechanical behavior of tubular joints, Doctoral dissertation, Osaka University, January 1967 (in Japanese).

4.40  Noel, J.S., Beale, L.A. and Toprac, A., An investigation of stresses in welded T-joints, Technical Report No.P.550-3, University of Texas, Austin, March 1965.

4.41  Beale, L.A. and Toprac, A.A., Analysis of in-plane T, Y and X welded tubular connections, Welding Research Council Bulletin, October 1967.

4.42  Andian, L.E., Sewell, K.A. and Womack, W.R., Partial investigation of directly loaded pipe T-joints, Southern Methodist University of Dallas, 1958.

4.43  Toprac, A.A., An investigation of welded steel pipe connections, Welding Research Council Bulletin, No.71, August 1961.

4.44  Toprac, A.A. and Louis, B.J., Research in tubular joints, Static and fatigue loads, OTC, Paper No.1062, 1969.

4.45  Washio, K., Togo, T. and Mitsui, Y., Experimental study on local failure of chords in tubular truss joints, Part I, Technology Report, Osaka University, Vol.18, pp.559-581, October 1968.

4.46  Washio, K. and Kurobane, Y., Truss joints in tubular steel structures (Preliminary Report), Technology Report, Osaka University, Vol.13, No.553, 1963.

4.47  Washio, K. and Mitsui, Y., High stress fatigue tests of tubular T-joints, Summary Papers, Annual AIJ Conference, August 1969 (in Japanese).

4.48   Wardenier, J. and Koning, C.H.M., Investigation into the static strength of welded Warren type joints made of circular hollow sections, TNO/IBBC Stevin Report B1-77-19, Stevin Laboratory, Delft University of Technology, July 1977.

4.49   Offshore Technology Conference, OTC, Paper No.3692, Texas, 1980.

4.50   Yamasaki, I., Takizama, S. and Komatsu, M., Static and fatigue tests on large-size tubular T-joints, OTC, Paper No.3424, 1979.

4.51   Yura, J.A., Howell, L.E. and Frack, K.H., Ultimate load tests on tubular connections, Civil Engineering Research Laboratory, Report No.78-1 to EXXON Production Company, University of Texas, Austin.

4.52   Zimmerman, W., Tests on panel point type joints of large diameter, Institut Otto Graf, Stuttgart, September 1965.

4.53   Study on tubular joints used for marine structures, The Society of Steel Construction of Japan, March 1972.

4.54   Fatigue strength of K, T and Y-joints in tubular structures, Research Institute of Ishikawajima-Harima Heavy Industries Co. Ltd., Unpublished Report, March 1978 (in Japanese).

4.55   Makino, Y., Kurobane, Y. and Minoda, Y., Strength of tubular X- and T-joints under tensile brace loading, Research Report No.5, Chugoky-Kyushu Branch of AIJ, March 1981 (in Japanese).

4.56   Takizawa, S., Yamamoto, N., Mihara, J. and Okata, S., Full-scale experiments of T- and X-type tubular joints under static and cyclic loading, Kawasaki Steel Technical Report, Vol.11, No.2, 1979, Kawasaki Steel Co. Ltd. (in Japanese).

4.57   Makino, Y., Kurobane, Y., Mitsui, Y. and Yasunaga, Y., Experimental study of ultimate strength of tubular joints with high strength steel and heavy walled chord, Research Report No.4, Chugoky-Kyushu Branch of AIJ, February 1978 (in Japanese).

4.58   Kaiho, Y., Akiloto, T., Kamiya, S. and Kamagoe, E., Study on structure of intersection of spherical tank pipe bracings, Kawasaki Steel Technical Report, No.64, August 1977, Kawasaki Steel Co. Ltd. (in Japanese).

4.59   Washio, K., Togo, T. and Mitsui, Y., Cross joints of tubular members, Report of Kinki Branch of AIJ, May 1966 (in Japanese).

4.60   Sparrow, K.D., Ultimate strengths of welded joints in tubular steel structures (taken from J. Wardenier, see Ref. 4.78).

4.61   Wardenier, J. and de Koning, C.H.M., Investigation into the static strength of welded joints with three bracings made of RHS on CHS, TNO Report BI-77-37/35.3.51210, Stevin Report 6.77.6.

4.62   Wardenier, J. and de Koning, C.H.M., The static strength of welded CHS K-joints, Stevin Report 6-81-13, TNO/IBBC Report B1-81-35/63.5.5470.

4.63   Toprac, A.A., Johnson, L.P. and Noel, J., Welded tubular connections, On investigation of stresses in T-joints, Welding Journal, Vol.45, No.1, January 1966.

4.64   Toprac, A.A., Natarajan, M., Erzurumlu, H. and Kanoo, A.L.J., Research in tubular joints, Static and fatigue loads, OTC, paper No.1062, 1969.


**Publications providing a synthesis of static strength tests**

4.65   Billington, C.J., Lalani, M. and Tebbett, I.E., Background to new formulae for the ultimate limit state of tubular joints, OTC, Paper No.4189, 1982.

4.66   Kurobane, Y., Makino, Y. and Mitsui, Y., Ultimate strength formulae for simple tubular joints, IIW Doc.XV-385-76, May 1976, Department of Architecture, Faculty of Engineering, Kumamoto University.

4.67   Kurobane, Y., Makino, Y. and Mitsui, Y., Re-analysis of ultimate strength data for truss connections in circular hollow sections, IIW Doc.XV-461-80, Faculty of Engineering, Kumamoto University.

4.68   Pan, R.B., Plummer, F.B. and Kuang, J.G., Ultimate strength of tubular joints, OTC, Paper No.2644, 1976.

4.69   Petit, L., Le comportement et la résistance des assemblages soudés de profils creux, soumis à des charges statiques, Monograph No.6, Section 6 Assemblage de profils creux circulaires avec découpes d'intersection, COMETUBE, 1979.

4.70   Rodabaugh, E.C., Review of data relevant to the design of joints for use in fixed offshore platforms, WRCB, 1980.

4.71   Trezos, C., Etude probabiliste de la résistance ultime des assemblages soudés en K, Construction métallique, 1, 1978.

4.72 Washio, K., Toko, T. and Mitsui, Y., Experimental study on local failure of chords in tubular truss joints, Part I, Technology Reports of the Osaka University, No.18, pp.559-581, October 1968.

4.73 Yura, J.A., Zettlemoyer, N. and Edwards, I.F., Ultimate capacity equations for tubular joints, OTC, paper No.3690, 1980.

4.74 Report on the UEG project definition study on design guidarce on tubular joints, CIRIA/UEG, Vols.I and II, May 1980.

4.75 Kurobane, Y., New developments and practices in tubular joint design, Faculty of Engineering, Kumamoto University, IIW Doc.XV-488-81, XIII-1004-81, May 1981.

4.76 Kurobane, Y., Bases for design of tube-to-tube joints with circular hollow sections, Addendum to New developments and practices in tubular joints design, Faculty of Engineering, Kumamoto University, Addendum to IIW Doc.XV-487-81, XIII-1004-81, August 1981.

4.77 Makono, Y., Kurobane, Y. and Minoda, Y., Design of CHS X- and T-joints under tensile brace loading, Faculty of Engineering, Kumamoto University, IIW Doc.XV-487-81, May 1981.

4.78 Wardenier, J., Hollow Section Joints, Chap.IV, Delft University Press, 1982.

4.79 Goyet, J., CTICM Report No.9001-3, July 1983 (Confidential).

# PART III

# FATIGUE ANALYSIS
# OF TUBULAR JOINTS

CHAPTER **1**

# Concepts

## 1.1 FATIGUE ANALYSIS OF TUBULAR JOINTS

The fatigue analysis of welded tubular joints requires the use of suitable S-N curves, which relate the design stress range to the number of cycles characterizing failure. The concept of stress range is defined in Chapter 2. A precise definition of failure is given in Chapter 6.

*The design stress range is the stress range that is plotted on the ordinate of the appropriate S-N curve for fatigue analysis.*

## 1.2  DEFINITION OF DESIGN STRESS AND STRESS CONCENTRATION FACTOR FOR A GIVEN LOADING

For a given loading, the design stress ($\sigma_G$) corresponds to the maximum stress at the weld toe, on the chord side or on the brace side. It only takes account of the global joint geometry. For a simple load, this design stress ($\sigma_G$) is the product of the nominal stress ($\sigma_n$), obtained by the standard structural analysis methods, and of the geometric stress concentration factor (SCF).

Hence

$$\sigma_G = \sigma_n \times K_G$$

$$SCF = K_G$$

The point where $\sigma_G$ occurs is called the "hot spot" of the joint for the load concerned.

The stress concentration factor can be determined with varying difficulty according to the case examined, either by calculation or by test (see Chapter 3).

By convention, the nominal stress is the stress applied to the welded tube (brace). Thus a nominal stress is associated with each brace-chord junction. Consequently the stress concentration factor is in turn associated with this junction (see Section 3).

**The procedure for determining nominal stresses**

For a given load case:

1. Determine the stresses at the member ends by linear elastic structural analysis. The structural joints are represented by the inter-sections of the centroidal axes (Fig. 1.1).

2. Using the reference of the member (brace E in Fig. 1.2) retain only the nominal loads $M_y$, $M_z$ and $N_x$ in the final calculation of the design stress, the other loads $M_x$, $N_y$ and $N_z$ being ignored.

By less conservative assumptions, using the design loads thus selected and the knowledge of the external loads directly applied to the member, one can calculate the loads related to the imaginary connection surface (Fig. 1.2).

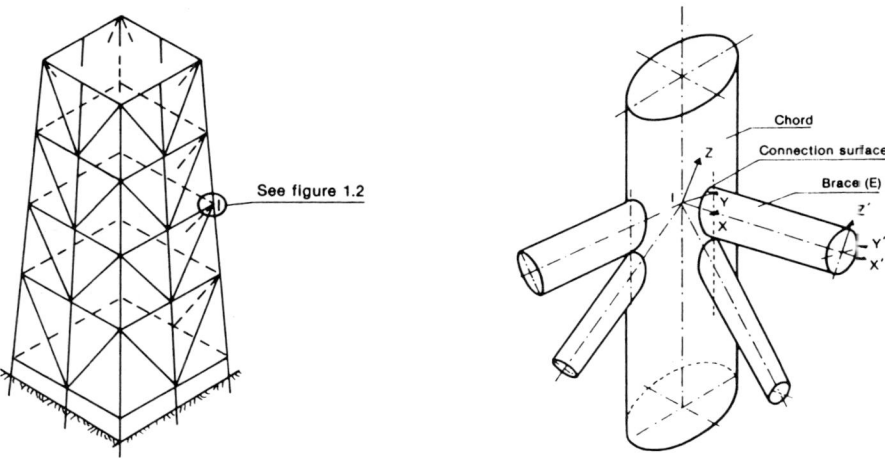

Fig. 1.1.                                      Fig. 1.2.

3. For each of the loads retained ($M_y$, $M_z$, $N_x$) calculate the nominal stresses related to the tube cross-section, taking account of the elastic moment of inertia of the tube and of its cross-section, i.e. $\sigma_{FY}$, $\sigma_{FZ}$ and $\sigma_{AX}$ (see Fig. 1.3).

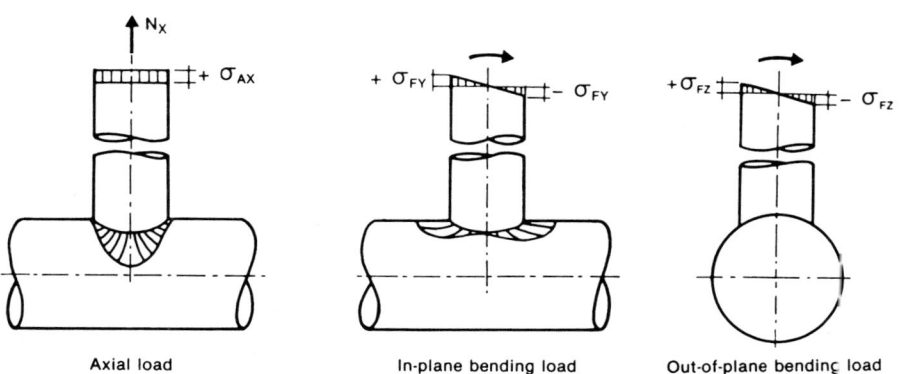

Axial load          In-plane bending load          Out-of-plane bending load

Fig. 1.3.

*The value of the local stress ($\sigma_L$) at the weld toe depends on the shape of the weld, the intersection profile between the weld and the outer wall of the chord or the brace, and, finally, very local notch effects. This local*

stress is very difficult to determine either
experimentally or by calculation. Moreover, this stress,
by its very nature has a random value. Therefore the idea
of trying to define the design stress by this local stress
was discarded.

It is obvious that the value of this local stress
partly conditions the joint fatigue life. This local
stress affects the crack initiation period in particular.

To limit the influence of the "local geometry"
parameter and to avoid incurring the risk of premature
cracking due to poor local geometry (or to a local
defect), Section 2.2, Part I, precisely defines the
conditions to be met for the weld toe intersection.

Given the random character of the influence of local
effects on the value of the local stress, the design
stress $\sigma_G$ is adopted, which depends only on the joint
geometry and its loading. This stress provided the basis
for plotting the conventional S-N curves [1.1].

For a given load, the point at which the highest stress
value is located is called the "hot spot" of the joint.
The hot spot changes position according to the joint
geometry and its loading. In joints with symmetrical
geometry that are symmetrically loaded, at least one pair
of hot spots exists per loading. In the remainder of this
guide, hot spots will be treated in the singular.

As a rule, hot spots are found on the outer skin of the
chord. For a simple loading, and depending on the loading,
the hot spot may be either in the neighborhood of the
saddle point or of the crown point (Fig. 1.4).

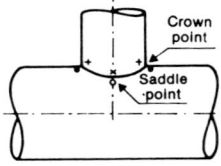

| Element | Brace loading type | | |
|---------|-------------|---|---|
| | Axial load | In plane bending | Out of plane bending |
| Chord | 0 | ● | 0 |
| Brace | × | + | × |

Hot spot location

Fig. 1.4.

For a complex loading, the position of the hot spot can
be determined with varying difficulty depending on the
analytical method employed (see Chapter 3). In a complex
joint, for example, with several braces, the number of
nominal stresses is equal to the number of welded tubes
(braces).

Figures 1.5 and 1.6 illustrate the design stress (or
geometric stress) at the weld toe of a brace-chord
junction, for the case of the T joint under axial load in
the brace. The stress distribution resulting around the
weld on the chord size and on the brace side is indicated
in Fig. 1.7.

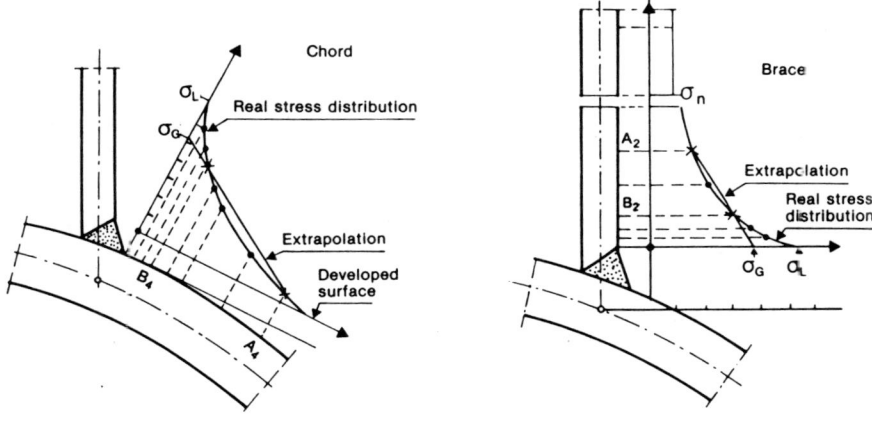

Fig. 1.5.                          Fig. 1.6.

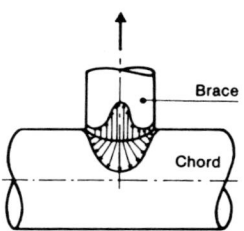

Fig. 1.7.

## 1.3  DEFINITION OF THE DESIGN STRESS RANGE

When the fatigue strength associated with a particular weld is evaluated, the geometric stress range (with time) calculated from the nominal stress range is taken into account. It can be expressed as follows:

$$\Delta\sigma_G = \Delta\sigma_{AX} \cdot SCF'_{AX} + \Delta\sigma_{FY} \cdot SCF'_{FY} + \Delta\sigma_{FZ} \cdot SCF'_{FZ}$$

$SCF'_{AX}$ = The value of the SCF for an axial load $N_X$ calculated at the hot spot corresponding to complex loading,

$SCF'_{FY}$ = The value of the SCF for the in-plane bending load $M_Y$, at the hot spot corresponding to the complex loading,

$SCF'_{FZ}$ = The value of the SCF for the out-of-plane bending load $M_Z$, at the spot corresponding to the complex loading.

*The value of $SCF'_{AX}$, $SCF'_{FY}$, $SCF'_{FZ}$ are not generally known at each point of the weld toe, nor at the hot spot position under complex loading. In this case, one can take known values of the SCF at the hot spots corresponding to simple loadings (see Section 3.2.3) and the formula that gives $\Delta\sigma_G$ takes on a symbolic character. This is because in principle, it is incorrect to add maximum stresses which do not necessarily occur at the same point. Section 3.2.3 gives the argument underlying the scalar and additive character of the terms that are added to each other to evaluate $\Delta\sigma_G$.*

## REFERENCE

1.1  Radenkovic, D., Analysis of stresses in tubular joints, Plenary
     Session 1, ECSC/IRSID International Conference, Steel in Marine
     Structures, Paris, October 1981.

CHAPTER **2**

# Actions and Loads

The use of S-N curves to evaluate the damage sustained by welded tubular joints requires knowledge of the stress history of the joint. Chapter 3 shows how to assess the design stress ranges as a function of the nominal stress ranges. The subject of this Section is the calculation of the nominal stress ranges with time, which will serve to develop the histogram of stress ranges for the subsequent damage calculations.

The history of a stress $\sigma(t)$ (axial or bending) (Fig. 2.1) at a pcint of the structure helps to determine:

(a) The value $\sigma_i$ at the peak of the ith cycle.

(b) The value $\sigma_i'$ at the trough of the ith cycle.

(c) The value $\Delta\sigma_i = \sigma_i - \sigma_i'$.

(d) The mean value $\sigma_i^m = \frac{1}{2}(\sigma_i + \sigma_i')$.

$\sigma_i$ represents the amplitude of a peak, while $\sigma_i$ represents the range of the stress cycle. This stress range is taken into account to calculate the geometric stress range (the S-N curve used for the damage calculation is plotted as a function of this stress range).

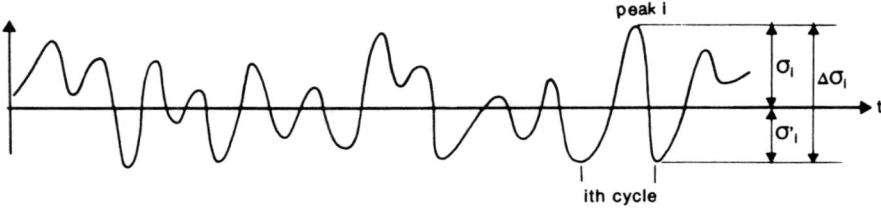

Fig. 2.1.

The load histogram is determined in 3 steps:

(1) Modelling of actions.

(2) Calculation of the frequency of occurrence of these actions.

(3) Calculation of loads in the structure subjected to the actions.

*The concept of stress cycle presents no difficulties of interpretation if the stress history reveals as many peaks (or troughs) as passages through the mean stress. In some cases (for example, if the stress history represents a trajectory of a wide band random process (see Section 2.1.4.B)), there are far more peaks (or troughs) than passages through the mean value, so that a numerical counting method must be resorted to. Various counting methods are available, and the choice between them will be discussed in Chapter 8 (cumulative damage).*

*The accuracy of the overall fatigue calculation depends on that of the different steps. This implies a degree of refinement (and generally an order of magnitude of the cost of computations) that is comparable for each step. Tables 2.1 and 2.2 summarise the compatibility of the different approaches.*

## 2.1  ACTIONS TO BE TAKEN INTO ACCOUNT

Any action producing stress variations in the joint under consideration during the phases of transport, installation and operation of the structure, must be taken into account in calculating damage.   The following must therefore be considered:

(a) Variable, cyclic and repetitive actions.

(b) Non-cyclic forces liable to alter the stresses induced by cyclic forces.

Environmental actions, which are cyclic or repetitive, liable to participate in the damage are:

(a) Waves:
   - Hydrodynamic drag and inertia forces.
   - Variations in hydrostatic thrust in the splash zone, wave breaking, slamming.

(b) Wind (dynamic forces).

(c) Ice impact.

*When the current is superimposed on the wave, the stress range in the joint may be altered.   However, consideration of this mechanism in calculations raises delicate problems.   It is routine practice to calculate separately the damage due to wave and the damage due to currents (vibrations generated by Von Karman eddies [2.1], reversal of tidal currents).*

## 2.1.1  HYDROSTATIC FORCES, SLAMMING, WAVE BREAKING

The structural members located in the neigborhood of the free surface are subject to variations in hydrostatic thrust due to the fluctuation in the elevation of the free surface, and may even be free of the water surface.   This mechanism participates in the damage.

The force due to slamming can be calculated as a drag force, and the drag coefficient is then replaced by a slamming coefficient.   This coefficient is often estimated from test results.   Moreover, the dynamic nature of the mechanism requires consideration of an amplification factor in evaluating these   loads   and,  finally,  the  frequency  of  the mechanism must be estimated [2.1, 2.10]. The same approach can be used for wave breaking.

*Fatigue analysis must be conducted on joints located in the plane of a conductor pipe support grid situated in the splash zone.*

*Slamming occurs on members of the structure which are slightly tilted to horizontal and located in the neighborhood of the free surface.*

*The breaking wave induces forces that may be several times greater than those determined by standard wave theories. It is mainly the vertical members that are sensitive to this factor.*

## 2.1.2 WIND

Wind is described by:

(a) The mean speed, which fluctuates slightly.

(b) Rapid fluctuations in speed about the mean speed.

For certain floating structures, the wind generates a dynamic excitation. For fatigue calculations, a spectral approach can be employed.

*As a rule, dynamic loads due to wind are slight in comparison with those due to waves. The modelling adopted must be based on the site data.*

## 2.1.3 ICE

In certain geographic areas, stresses induced by ice impact should be taken into account, together with the frequency of this impact.

## 2.1.4 WAVE ACTION IN SUBMERGED STRUCTURES

Forces acting on the submerged parts of the structure are calculated by one of the two following methods:

(1) Integration of the fluid pressure, based on the calculated potential and using the diffraction theory.

(2) Morison formula, based on the kinematics of the undisturbed fluid and experimental coefficients [2.1 to 2.6].

The choice between these two approaches is based on the comparison of the dimensions of the structural members concerned and the wavelength. To apply Morison's formulation, the criterion often used is $\Lambda/D > 8$ in which $\Lambda$ is the wavelength and D is the smallest dimension of the member.

For fatigue calculations, one is interested in the stress variations at the point of the structure concerned, over a long time interval (reference period). Hence the fluid kenematics must be modelled as a function of time. Available long-term statistics help to break down the overall wave data into individual waves, which are modelled by periodic deterministic waves (the deterministic approach), or else into "short-term sea states", which are modelled by a stationary process with certain properties (the random approach). The choice between these two approaches largely depends on:

(a) Statistical data available (long-term statistics) covering:
. Periods.
. Heights.
. Propagation directions.

(b) The method envisaged for load calculations.

*The scatter relation associating the wavelength $\Lambda$ to the period T is indicated in Section 2.1.4C.*

*The reference period selected serves to assess the significance of the damage calculated. This concept is discussed in detail in Section 8 (cumulative damage). The period must also be sufficiently long for the statistical sampling of wave action over this interval to be realistic. This is why some regulations recommend that they should not be less than 20 years [2.1, 2.6].*

*The order of magnitude of the number of stress cycles over the reference period is typically $10^8$.*

*The stationarity of short-term sea states is observed over a time interval of a few hours. This period is short in comparison with the reference period, but long in comparison with the mean period of stress cycles.*

## A. Deterministic wave

The most widely used models for fatigue calculations are:

(a) Airy wave theory.

(b) Stokes wave theory, 5th order.

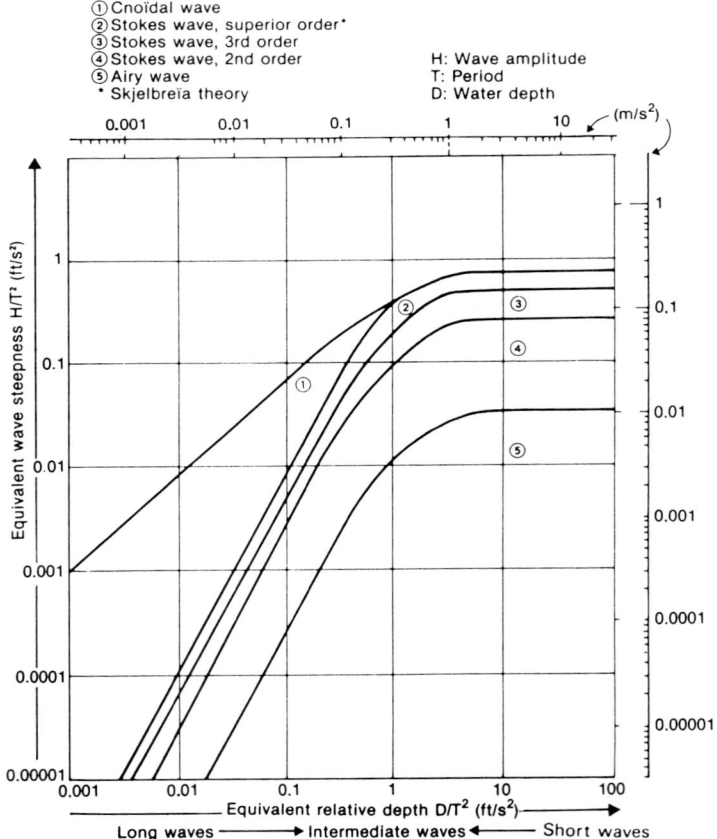

Fig. 2.2.

If the ratio of wave height to water depth is too high, the following models are more realistic:

(a) Cnoïdal wave theory.

(b) Current function theory.

The diagram in Fig. 2.2 shows the validity zones of these wave models.

> *The descriptive parameters are the propagation direction, period, crest-to-trough height, and water depth. The flow is assumed to be irrotational in a perfect incompressible fluid.*
>
> *The cnoïdal wave is used for shallow water situations.*
>
> *The Airy model, which is linear in relation to the crest-to-trough height, does not furnish values of the kinematics above the mean level of the free surface. For large wave heights, the Stokes model, for example, is more realistic at the free surface zone.*
>
> *In calculating the fluid acceleration, the convection terms (non-linear in relation to wave height H) may be non-negligible if H is not small compared to the wavelength Λ.*

## B. Random wave

The Gaussian model is the most widely used to describe short-term sea states. The spectral density function of the elevation of the free surface $W_\eta(\Theta, \omega)$ is defined, where $\Theta$ is the propagation direction and $\omega$ the circular frequency. $W_\eta$ suffices to characterise the short-term sea state.

The following is often adopted for an analytical representation of $W_\eta$:

$$W_\eta(\Theta, \omega) = f(\Theta) \; G_\eta(\omega)$$

where $f(\Theta)$ is the non-null direction function within an angular segment $|\Theta_1, \Theta_2|$. Among the most widely used are the following:

$$f(\Theta) = C \cos^4 (\Theta - \Theta_m) \quad \text{for } \Theta \text{ between} \quad \Theta_m - \frac{\pi}{2} \text{ and } \Theta_m + \frac{\pi}{2} \quad [2.18]$$

$$f(\Theta) = C \cos^2 (\Theta - \Theta_m) \quad \text{for } \Theta \text{ between} \quad \Theta_m - \frac{\pi}{2} \text{ and } \Theta_m + \frac{\pi}{2} \quad [2.19]$$

$$f(\Theta) = C \cos^4 (\frac{\Theta - \Theta_m}{2}) \quad \text{for } \Theta \text{ between} \quad \Theta_m - \pi \text{ and } \Theta_m + \pi \quad [2.20]$$

where C is a normalising constant and $\Theta_m$ the "mean direction" of the wave. It is generally in this direction that the wave energy is the greatest.

A routine simplification consists in defining a unidirectional spectrum ($f(\Theta) = \delta(\Theta - \Theta_m)$), where $\delta$ is the Dirac ditribution). $G(\omega)$ is then unidirectional spectral density function and $\sigma_\eta^2 = \int_0^\infty G_\eta(\omega)$. It must be confirmed that the use of the unidirectional spectrum ensures safety in comparison with the use of the directional spectrum for the joint concerned. The following developments concern a unidirectional spectrum. The spectral width parameter $\epsilon$ serves to assess the shape of the elevation of the free surface (Fig. 2.3).

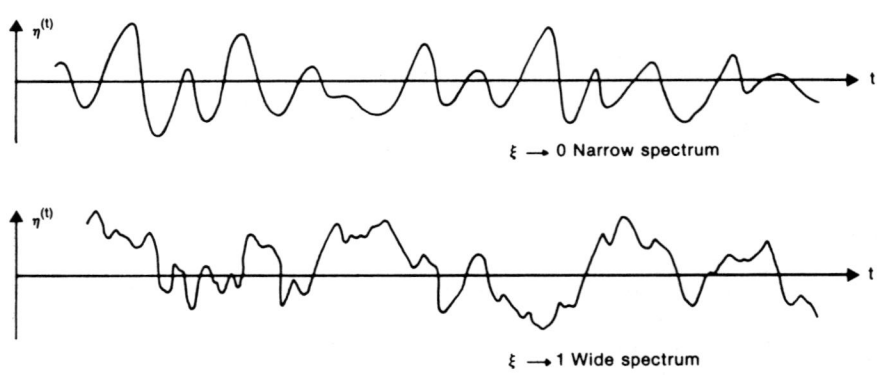

Fig. 2.3.

**Expression of the most usual unidirectional spectra:**

Pierson-Moskowitz:

$$G_\eta(\omega) = \frac{A}{\omega^5} e^{-\frac{B}{\omega^4}} \quad \left\{ \begin{array}{l} A = 0.0081 \ g^2 \\[6pt] B = 0.74 \ g^4 / W^4 \end{array} \right.$$

W = wind speed at a height of 18 m.

Modified expression of the Pierson-Moskowitz spectrum:

$$G_\eta(\omega) = \frac{A}{\omega^5} e^{-\frac{B}{\omega^4}}$$

$$A = \frac{\overline{H}^2_{1/3}}{4\pi}\left(\frac{2\pi}{\overline{T}_\eta}\right)^4$$

$$B = \frac{1}{\pi}\left(\frac{2\pi}{\overline{T}_\eta}\right)^4$$

Jonswap (Joint North Sea Wave Project):

$$G_\eta(\omega) = \frac{A}{\omega^5} e^{-\frac{B}{\omega^4}} \ \gamma^{\left[e^{-c(\omega)}\right]}$$

$$A = \frac{4\pi^3\,\overline{H}^2_{1/3}}{\overline{T}^4_\eta}$$

$$B = \frac{5}{4}\,\omega^4_m$$

$$C(\omega) = \frac{(\omega-\omega_m)^2}{2\sigma^2\omega^2_m} \qquad \sigma = \sigma_a \quad \text{if} \quad \omega \leq \omega_m$$

$$\sigma = \sigma_b \quad \text{if} \quad \omega > \omega_m$$

$\gamma$ varies between 1 and 7.

The Jonswap spectrum, the result of measurements conducted in the North Sea, is ideal for the sea during formation and near the coast.

*The elevation of free surfaces is denoted $\eta(x, y, t)$. The stochastic field $\eta(x, y, t)$ indexed to $R^3$ is homogeneous, centred, stationary in time, ergodic, Gaussian, of the 2nd order, and with a continuous quadratic mean.*

*Its spectral representation is written:*

$$\eta(x,y,t) = \int_0^{2\pi}\int_0^\infty e^{ik(\omega)\,[x\,\cos\Theta\,+\,y\,\sin\Theta]\,-\,i\omega t}\,dA_{\eta\eta}(\Theta,\omega)$$

$A_{\eta\eta}$ is the vectorial spectral process associated with $\eta(x,y,t)$, with a likelihood

$$\{|dA_{\eta\eta}(\Theta,\omega)|^2\} = \frac{1}{2} W_\eta(\Theta,\omega)d\Theta \; d\omega \quad for \quad \omega \geq 0$$

and where the variance of $\eta$ is:

$$\sigma_\eta^2 = \int_0^{2\pi}\int_0^\infty W_\eta(\Theta,\omega)d\Theta \; d\omega$$

The elevation of the free surface $\eta(x,y,t)$ is easily interpreted as the sum of an infinity of elementary (Airy) sinusoidal waves with circular frequency $\omega$ , direction $\Theta$ , of amplitude $\sqrt{W_\eta(\Theta,\omega)/2 \; d\Theta d\omega}$, and with a random phase uniformly distributed over $[\Theta,2\pi]$.       Hence wave trajectories can be generated by summation over a number of elementary waves.   This can also be done by smoothing a Gaussian white noise.

The estimation of the dependence of $W_\eta(\Theta,\omega)$ on $\Theta$ from site data is often delicate.

One has:

$$\int_{\Theta_1}^{\Theta_2} f(\Theta)d\Theta = 1$$

The spectral moments $(m_k = \int_0^\infty \omega^k \; G_\eta(\omega) \; d\omega)$ serve to express the following values concerning the elevation of the free surface:

(a) Mean frequency of passages through zero at positive steepness:

$$\overline{N}_o^+ = \frac{1}{2\pi} \sqrt{\frac{m_2}{m_o}}$$

(b) Mean frequency of peaks:

$$\overline{N}_m = \frac{1}{2\pi} \sqrt{\frac{m_4}{m_2}}$$

(c) Mean apparent period:

$$\overline{T}_\eta = 2\pi \sqrt{\frac{m_2}{m_4}}$$

*(d) Spectral width:*

$$\varepsilon = \sqrt{1 - \frac{m_2^2}{m_1 m_4}}$$

*The peak probability distribution is expressed by:*

$$f(x) = \frac{1}{\sqrt{2\pi m_0}} \left[ \varepsilon \, e^{-\frac{x^2}{2m_0 \varepsilon^2}} + \sqrt{1-\varepsilon^2} \, \frac{x}{\sqrt{m_0}} \, e^{-\frac{x^2}{2m_0}} \int_{-\infty}^{\frac{x\sqrt{1-\varepsilon^2}}{\varepsilon\sqrt{m_0}}} e^{-\frac{u^2}{2}} \, du \right]$$

*Narrow band spectrum:*

$$\varepsilon \longrightarrow 0 \quad f(x) \longrightarrow \frac{x}{m_0} \, e^{-\frac{x^2}{2m_0}} \qquad (Rayleigh's \ law)$$

*Wide band spectrum:*

$$\varepsilon \longrightarrow 1 \quad f(x) \longrightarrow \frac{1}{\sqrt{2\pi m_0}} \, e^{-\frac{x^2}{2m_0}} \qquad (Gauss's \ law)$$

As a rule, the real wave spectrum is not narrow. However, Rayleigh's law for the maxima is a good approximation of $f(x)$ for $\varepsilon < 0.5$. Similarly, if one can calculate the spectral density function of the stress at a joint (linear behavior of the structure in relation to the wave height and to nodal displacements), the width of the stress spectrum helps to define the stress range distribution, and hence to select the most suitable method for calculating cumulative damage.

For $\varepsilon < 0.5$, one can express:

(a) Mean trough $\overline{H} = \sqrt{2\pi m_0}$

(b) Significant height $\overline{H}_{1/3}$ (or $H_s$) $\simeq 4\sqrt{m_0}$.

More generally, we have:

$$\overline{H}_{1/3} = 2 \ C(\varepsilon)\sqrt{m_0}$$

*C( ε) varies as follows:*

| ε | 0.0 | 0.1 | 0.2 | 0.3 | 0.4 | 0.5 | 0.6 | 0.7 | 0.8 | 0.9 | 1.0 |
|---|-----|-----|-----|-----|-----|-----|-----|-----|-----|-----|-----|
| C( ε) | 2.00 | 2.00 | 1.99 | 1.98 | 1.96 | 1.93 | 1.89 | 1.83 | 1.74 | 1.57 | 1.09 |

$\overline{H}$ *is  defined  as  the  mean  of  the  crest-to-trough heights.*

$\overline{H}_{1/3}$ *is defined as the mean of the upper third of the crest-to-trough heights.*

### Modified Pierson-Moskowitz spectrum:

*Theoretically,  ε = 1.  In practice, one obtains ε = 0.425 provided that ω is limited to $\omega_l$ so that:*

$$\int_0^{\omega_l} G_\eta(\omega)\ d\omega = 0.90 \int_0^{\infty} G_\eta(\omega)\ d\omega$$

*This value of ε justifies a posteriori the expression of A and B as a fonction of $\overline{H}_{1/3}$ and $\overline{T}_\eta$ since we obtain $\overline{H}_{1/3} = 4\sqrt{m_o}$, which is correct for  ε < 0.5.*

*This  spectrum  closely  corresponds  to  reality  for  a formed sea.*

### JONSWAP spectrum:

*$\omega_m$ represents the circular frequency for which $G_\eta(\omega)$ is a maximum.*

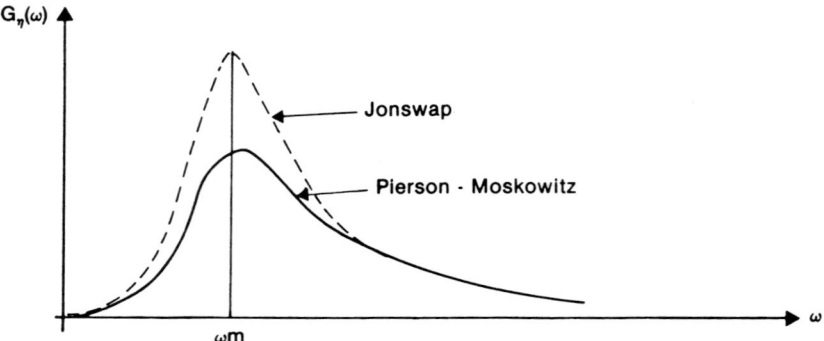

*Fig. 2.4.*

*γ is a dimensionless parameter representing the ratio of the peak of the JONSWAP spectrum to the peak of the Pierson-Moskowitz spectrum with the same $\bar{H}_{1/3}$ and $\bar{T}_\eta$, both peaks appearing at the same angular frequency (Fig. 2.4).*

## C. Consideration of the long-term

The data required concerning the long-term depend on the choice described above between the deterministic wave and the random wave. Apart from the description of individual waves by a periodic deterministic wave model, it is necessary to know the distribution over the reference interval of the height, period and propagation direction of the individual waves. Apart from the description of short-term sea states by a random wave model, it is necessary to know the long-term distribution of the short-term parameters (in general $\bar{H}_{1/3}$, $\bar{T}_\eta$ and $\Theta_m$ [2.11, 2.21, 2.22]).

*The short-term description whether deterministic or random, depends on the choice of the long-term approach, which itself depends on the available site data.*

### Random wave

Complete knowledge of the waves over the reference period is obtained from the data of the three dimensional joint law of short-term parameters $\bar{H}_{1/3}$, $\bar{T}_\eta$ and $\Theta_m$. These data are not always available for the North Sea regions or elsewhere. In addition, many records deal with the "visual" parameters $H_v$ and $T_v$.

Although, $\bar{T}_\eta$ is a random variable, the following formulas can be used for fixed $T_v$ [2.1, 2.12]:

$$\bar{H}_{1/3} = 1.68 \, H_v^{0.75}$$

$$\bar{T}_\eta = 0.82 \, T_v^{0.96}$$

To use an analytical expression of the distribution law $\bar{H}_{1/3}$, one can adopt a three-parameters Weibull law [2.13]:

$$P(\bar{H}_{1/3} \leq x) = 1 - \exp\left\{-\left(\frac{x-H_o}{H_c-H_o}\right)^\gamma\right\}$$

This makes it necessary to calculate $H_o$, $H_c$ and $\gamma$ from site data. It is often assumed that $H_o = 0$, leaving two parameters.

The conditional law of $\overline{T}_\eta$ to $\overline{H}_{1/3}$ given is also represented by a two-parameter Weibull law:

$$P(\overline{T}_\eta \leq Y/\overline{H}_{1/3} = x) = 1 - \exp\left\{-\left(\frac{Y}{T_c(x)}\right)^{\gamma(x)}\right\}$$

where $\gamma(x)$ and $T_c(x)$ are functions of the value $x$ of $\overline{H}_{1/3}$ [2.16].

*Examples can be provided of data for the North Sea (60°N, 02°E) compiled from 1961 to 1975, which give:*

$$P(\overline{H}_{1/3} \leq x) = 1 - \left\{\exp\ -\left(\frac{x}{2.298}\right)^{1.520}\right\}$$

February 1973 - February 1975

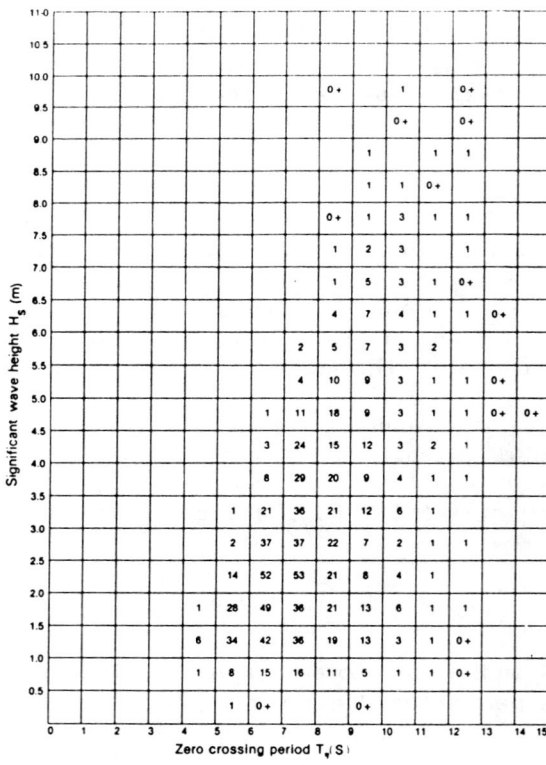

Scatter diagramm corresponding to 1 000 short term
sea states (0 + signifies less than 0.5)

*Fig. 2.5.*

*The joint probability laws of $\bar{H}_{1/3}$ and $\bar{T}_\eta$ can be represented in the form of scatter diagrams which indicate the frequencies of occurrence of each pair ($\bar{H}_{1/3}$, $\bar{T}_\eta$) (Fig. 2.5).*

**Individual waves**

Two approaches can be distinguished:

(1) Statistical approach.

The ideal case is one in which records are available over a sufficiently long time interval, for the height and period of individual waves, as well as their propagation direction.

*In general, long-term records of individual wave heights and periods only cover a few years of observation. Therefore the distribution law derived is very likely to be incorrect for large wave heights. This state of affairs is less serious for fatigue problems, in which large waves are not the most determinant in cumulative damage calculations than for other problems, such as the determination of the "project design wave" (Chapter 2, Part II).*

*Furthermore, it is always difficult to estimate the propagation direction.*

(2) Probabilistic approach.

This consists in combining the distribution of wave heights for a sea state characterized by $\bar{H}_{1/3}$ and $\bar{T}_\eta$ with the distribution of $\bar{H}_{1/3}$ and $\bar{T}_\eta$ over the reference period. This gives the distribution of the individual wave heights over the reference period. In certain conditions, the result is close to a Weibull distribution [2.13].

$$P(H > H^*) = \exp\left\{-\left(\frac{H^*}{H_1}\right)^\gamma\right\}$$

where $H_1$ and $\gamma$ are constants related to the site.

*This result, obtained by Nordenström [ 2.13] assumes
that short-term sea states are characterised by narrow
band spectra (ε approaching 0).*

If $\gamma = 1$, the log-linear distribution is obtained, which is routinely used in practice for the North Sea:

$$\ln(N) = \ln(N_o) - \frac{H^*}{H_1}$$

where N is the number of waves with height greater than H*, and $N_0$ is the total number of waves.

This approach does not allow to associate a period with the wave height thus determined. A realistic period can be found by considering the wave camber (ratio $H/\Lambda$ , where H is the height and $\Lambda$ the wavelength. The camber is typically 1/15. As a rule, a camber near the upper theoretical limit is selected (corresponding to wave breaking). However, more severe conditions can be obtained if one selects a wave period close to the natural frequency of the structure, or corresponding to a maximum hydrodynamic excitation, arising from the geometrical configuration of the structure.

Figure 2.6 gives examples of maximum cambers for individual waves at a North Sea location.

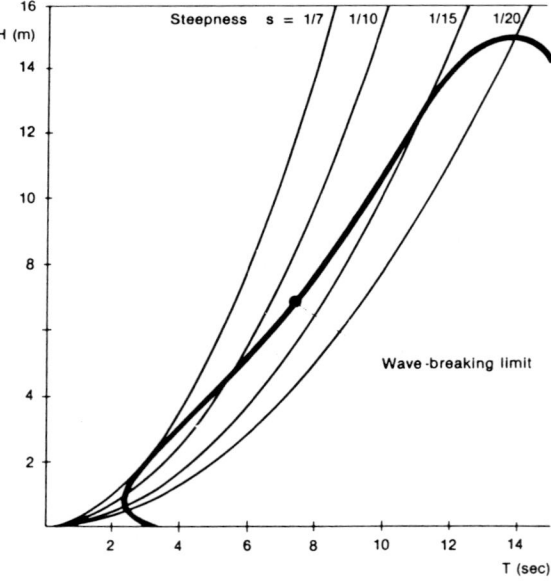

Fig. 2.6.

*The wavelength $\Lambda$ and the period $T$ are connected by the scatter equation involving the depth $D$:*

$$\frac{\Lambda}{T^2} = \frac{g}{2\pi} \tanh \frac{2\pi D}{\Lambda}$$

*$g$ = gravitational acceleration.*

*At extreme depth ($\tanh (\frac{2\pi D}{\Lambda})$ approaching $1$):*

*$\Lambda = 1.56 \ T^2$ ($\Lambda$ en m, $T$ in s).*

*Studies are available concerning the joint height-period distribution relation for a given short-term sea state [ 2.14, 2.15 ]. The knowledge of such a relation helps to calculate a joint height-period distribution law for individual waves over the reference period, and hence to associate a height $H$ with a period $T$ for a given probability of occurrence.*

## D. Development of a stress concentration histogram

When the wave is described by short-term sea states, the stress range histogram results either from the application of a cycle counting technique to the results of a simulation, or from an analytical calculation after the application of the spectral method (if the process can be considered to be of narrow band width). At least ten short-term sea states need to be considered.

When the wave is described by individual waves for which a histogram of heights and a height-period scatter diagram are generally available, the stress range histogram is obtained point by point, each point requiring the analysis of the stresses in the structure subject to a given wave. The following guidelines can be provided on the number of these calculations likely to yield a final result of sufficient accuracy:

(a) If dynamic effects are slight, at least four design waves must be considered per propagation direction analysed. The heights of these waves must be distributed in the range of heights corresponding to the site, with due consideration given to the fact that damage essentially originates in the moderate wave height zone.

(b) If dynamic effects are present in the structure's response to the wave, the stress range histogram is different locally from that of the "static" stress range, i.e. without dynamic effects. The number of waves to be considered must therefore be increased, by selecting additional wave periods in the resonance zone.

## 2.2  LOAD CALCULATIONS

The knowledge of member forces (moments, axial and shear forces) requires calculation of the displacements of structural joints under the action of external forces applied directly to the joints and to the discretised elements of the structure. Once the nodal displacements are known, the nominal stresses are calculated by considering each member of the joint individually.

For a linear elastic calculation, if F is the vector representing the external forces applied to the joints of the structure, the vector X of nodal displacements is the solution to:

$$KX = F(X, \dot{X}, \ddot{X}, t, D, \dot{D}, \ddot{D})$$

where K is the stiffness matrix of the structure, t the time, and D the overall displacement vector of the structure. F is obtained by modelling the forces, as discussed in Section 2.1. Its form closely depends on the type of structure considered.

Fixed structures: Overall displacements of the structure are nil, hence $F = F(X, \dot{X}, \ddot{X}, t)$. If the fluid-structure interaction is ignored, F is written: $-M \ddot{X} - C\dot{X} + F'(t)$.

Mobile structures: overall displacements of the structure occur. In most cases, X is supplanted by D to calculate $F = F(t, D, \dot{D}, \ddot{D})$. Calculation is carried out in three steps:

(a) Calculation of D, $\dot{D}$, $\ddot{D}$ assuming the structure to be rigid (X=0).

(b) Calculation for any value of t, $F'(t) = F(t, D, \dot{D}, \ddot{D})$.

(c) Statistic analysis of the structure: solution of $KX = F'(t)$.

From this standpoint, Tables 2.1 and 2.2 present the overall fatigue calculation for these two types of structure.

The differential equation that has to be solved is finally one of the following three types:

(1) Quasi-static behavior $KX = F(t)$.

(2) Linear behavior $M\ddot{X} + C\dot{X} + KX = F(t)$.

(3) Non-linear behavior $KX = F(t, X, \dot{X}, \ddot{X})$.

## 2.2.1 CONSTRUCTION OF LOAD CASES

### Static analysis

The load case consists of given external forces applied to the joints and to the discretized elements. Certain analysis programs contain input modules for the automatic generation of such load cases, starting from the wave, for a fixed structure.

### Time domain analysis

The load cases are obtained by carrying out a simulation of the wave for fatigue calculation, i.e. by generating the fluid kinematics for each calculation instant, with the external forces calculated from these kinematics.

### Frequency domain analysis

In this case, the wave is known by its crest-to-trough height, period and direction. The load case is given by the amplitude of each nodal force component and its phase displacement in relation to the wave.

## 2.2.2 DISCRETIZATION OF THE STRUCTURE

For the fatigue analysis of a tubular joint, the stresses in the members of the assembly concerned must be estimated as accurately as possible. This requires a certain precision in the degree of discretization of the structure. A highly simplified model is liable to lead to substantial errors (in the discussion which follows, n denotes the number of degrees of freedom of the model).

*For example, it is possible, with a simplified model, to correctly estimate the horizontal displacements of the various horizontal levels of the "jacket". For certain joints however, this is not sufficient to calculate the stresses accurately.*

## 2.2.3 LINEAR EQUATION

Nodal displacements are solutions to the equation:

$$M\ddot{X} + C\dot{X} + KX = F(t)$$

where X and F(t) are column-vectors of dimension n.

### A. Characteristics modes of undamped systems

These are non-null solutions to:

$$-\omega^2 MX + KX = 0$$

Each natural mode is associated with a natural period $T_i = \dfrac{2\pi}{\omega_i}$ for which $\det [K - \omega_i^2 M] = 0$.

The value of the basic natural period $T_0 = \sup T_i$ serves to determine whether the structure's response to the wave will be quasi-static ($T_0$ much lower than the wave periods), or whether the dynamic effects will enter into action ($T_0$ close to the wave periods). If the structure is sufficiently rigid, the following static approximation may suffice:

$$X(t) = K^{-1} F(t) = X_{st}(t)$$

If not, it will be necessary to solve the dynamic equation which supplies the nodal displacement $X_{dyn}(t)$. For a harmonic loading F(t) with angular frequency one can write:

$$\left| X_{dyn}^i(t) \right| = K_i(\omega) \left| X_{st}^i(t) \right| \qquad i = 1 \text{ to } n$$

where $X_{dyn}^i$ and $X_{st}^i$ are the dynamic responses according to the modes i and $K_i(\omega)$ is the dynamic amplification factor for mode i.

*For a fixed "jacket" type structure, the natural period is generally less than three seconds. However, it has been observed that the natural period tends to increase with water depth (i.e. the height of the "jacket").*

### B. Static calculations

The calculation must be repeated for all the calculation instants, the stiffness matrix having been inversed only once at the outset. The case of the static calculation of a fixed structure under deterministic waves is described in detail in Section 2.2.5.

### C. Dynamic calculations

The three best known methods are presented below:

(1) Frequency resolution in spatial coordinates:

If F(t) is broken down into a Fourier series (coefficient $C_F(\omega)$, of dimension n), the solution vector X is given as the sum of a Fourier series whose coefficients $C_X(\omega)$, of dimension n are as follows:

$$\{-\omega^2 M - i\omega C + K\} \, C_x(\omega) = C_F(\omega)$$

Hence the matrix must be inversed in order to obtain a coefficient.

*This method can be used if the excitation is periodic and deterministic. It can also serve to calculate the nodal force/nodal displacement transfer function, and hence serves as a basis for spectral calculations (see Section 2.2.3D). It offers the advantage of being precise (in accordance with the linear behavior of the structure). In particular, the calculated response includes local displacements, as opposed to displacements involving the entire structure, which is important for calculations of the loads in a joint. However, like all resolution methods in spatial coordinates, it raises the problem of the construction of the damping matrix. As a rule, since damping is defined by a critical damping rate for each mode, one must at first extract the natural modes in order to calculate the damping matrix.*

(2) Pure modal superposition.

The characteristic modes presented in Section 2.2.3A are mutually orthogonal. Hence it is important to rewrite the nodal displacement equation on the basis of the natural modes.

This gives:

$$X = \sum_i \Phi_i Y_i$$

where $\Phi_i$ are vectors representing the characteristic modes ($i = 1$ to $n$) and $Y_i$ the modal coordinates for nodal displacements. The new unknowns $Y_i$ are solutions to the system of $n$ equations:

$$\sum_j \left\{ {}^t\Phi_i M \Phi_j \ddot{Y}_j + {}^t\Phi_i C \Phi_j \dot{Y}_j + {}^t\Phi_i K\Phi_j Y_j \right\} =$$

$$= {}^t\Phi_i F(t), \qquad i = 1 \text{ to } n$$

and the orthogonality of the mode:

$${}^t\Phi_i M \Phi_i \ddot{Y}_i + \sum_j {}^t\Phi_i C \Phi_j \dot{Y}_j + {}^t\Phi_i K \Phi_j =$$

$$= {}^t\Phi_i F(t), \qquad i = 1 \text{ to } n$$

The method is interesting if the matrix C is such that $\Phi_i C \Phi_j = 0$ when $j \neq i$. One then obtains $n$ decoupled linear equations, whose resolution is trivial. This situation prevails, for example, if damping is expressed in terms of critical damping rate per mode, which is routine practice.

*This method requires a truncation of the sum $X = \sum_i \Phi_i Y_i$. In fact, the extraction of natural modes is rather costly.*

*This is not a problem if one considers the overall behavior of the structure, which is generally governed by the response according to the first natural modes. However, the stress at a specific joint also depends on the responses according to very "local" modes. Hences the use of this method for accurate fatigue calculations [2.27] is a delicate matter. It applies to any loading F(t).*

(3) Modal superposition + static response.

The pure superposition method is improved at little expense by taking account of the static contribution of the response according to modes not taken into account dynamically. However, this method, like the previous one, only applies if a small number of modes are excited dynamically by $F(t)$.

*The foregoing modal superposition method gives the following dynamic contribution for the M characteristic modes taken into account:*

$$Y_i(t) = K_i(\omega) \, Y_i^S(t), \qquad i = 1 \text{ to } M$$

*where $Y_i^S(t)$ is the static contribution of mode $i$ and $K_i(\omega)$ the dynamic amplification coefficient for modes $i$ at the excitation circular frequency $\omega$. If the complete static response of the structure, or $X^S(t)$ is also calculated, an approximation of the complete dynamic response is obtained by:*

$$X(t) = X^S(t) + \sum_{i=1}^{M} (K_i(\omega)-1)\Phi_i Y_i^S(t) =$$

$$= X^S(t) + \sum_{i=1}^{M} (1 - \frac{1}{K_i(\omega)}) \, \Phi_i Y_i(t)$$

## D. Use of the transfer function

The foregoing methods are used to calculate the free surface elevation-joint stress amplitude transfer function. This transfer function can be used:

(a) In deterministic calculations.

(b) In stochastic calculations, to yield the spectral density function of the stress by:

$$G_\sigma(\omega) = \left| TF(\omega) \right|^2 G_\eta(\omega)$$

The methods discussed in Chapter 8 are then used to calculate the damage.

*This approach assumes that the excitation force $F(t)$ is linear in relation to the wave height. This restriction is sometimes discarded and a pseudo-transfer function calculated with each elementary wave modelled by a* *realistic height as a function of its period and of the location of the structure.*

## 2.2.4 NONLINEAR EQUATION

As one can no longer speak of transfer functions in this case. Two alternatives are available:

(a) Linearisation of the equation (replace the nonlinear terms by "equivalent" linear terms).

(b) Solution of the equation by time simulation.

*Among the main sources of nonlinearities are the following:*

*(a) Hydrodynamic damping of the drag calculated by the Morison equation, due to the velocity of the structure, especially the lattice members.*

*(b) The nonlinearity of the drag force applied to a member as a function of the fluid velocity.*

*(c) Consideration of the deformation of the structure in calculating the external forces (generation of the kinematics in the exact position of the structure).*

*(d) Nonlinear behavior of the foundation soil for a fixed structure.*

*(e) Consideration of the deformed free surface in calculating the hydrodynamic forces acting on the structure.*

## A. Equivalent linearisation technique

Let us consider the example of the nonlinear term of hydrodynamic drag:

$$F(X,\dot{X},t) = K_D \; |V-\dot{X}| \; (V-\dot{X})$$

where V is the fluid velocity.

This term is replaced by $A(V-\dot{X})$ in minimizing the error in each component in the least square sense. The coefficients of A depend on the result, and the method is hence generally iterative (it converges rapidly). At each iteration step, one of the methods described for the linear equation is employed. A discussion of this technique can be found in Ref.[2.28].

> When the linearised equation is treated by the modal superposition method, it is necessary to approximate the matrix A by a diagonal matrix in the modal base.

## B. Simulation techniques

This involves solving the nonlinear equation by numerical integration, with the excitation force calculated at each instant as a function of the nodal displacements at the previous time steps. A numerical integration algorithm is used, requiring at each time step the inversion of the matrices of rank n. It is also necessary to select an algorithm that guarantees numerical convergence.

One obtains a history of the response that is employed:

(a) To determine the stress range during a cycle, in the case of deterministic calculations.

(b) In applying a stress cycle counting method, in the case of stochastic calculations with a wide band response.

> The method can be used for stochastic calculations. In fact, based on a spectral density function of the free surface elevation, one can model a realisation of the process by summating a large number of sinusoidal elementary waves. One can also proceed by smoothing a Gaussian white noise. In any case the method is extremely costly.

## 2.2.5 SIMPLIFIED MODEL. DETERMINISTIC ANALYSIS

The method described here is used for its simplicity and low cost. It concerns fixed structures of the "jacket" type [2.30].

The prevailing framework, discussed in Section 2.1.4C, is that of the long-term modelling of the wave by a succession of individual waves.

The histogram of wave heights is determined from the site data. We shall consider here the case which assumes the log-linear law:

$$\ln(N) = \ln(N_o) - \frac{H}{H_o} \ln(N_o)$$

$N_o$ = total number of waves over the reference period,

$N$ = number of waves with height greater than H,

$H_o$ = most probable maximum height over the reference period (Fig. 2.7).

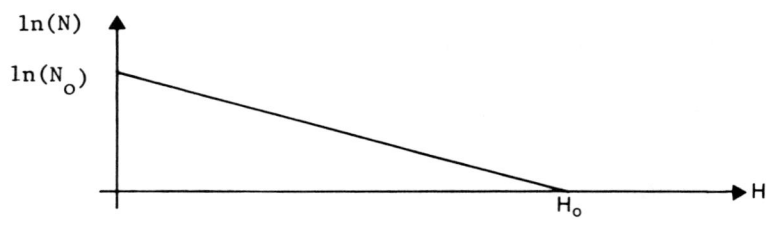

Fig. 2.7.

To obtain the histogram of nominal stress ranges in a joint, $\Delta\sigma$ must be associated with H. The first problem is to associate a period with each wave height. This is done by assuming that the wave camber is constant.

With this hypothesis, it is assumed that H and $\Delta\sigma$ are related by a simple analytical equation ($\Delta\sigma_s$ is the static stress range):

$$\Delta\sigma_s = \alpha \, H^\beta$$

$\alpha$ and $\beta$ are two constants which are determined from the results of stress calculations in the joint concerned for two wave heights. This gives the dynamic stress range in the joint considered by:

$$\Delta\sigma = K \, \Delta\sigma_s$$

The overall amplification coefficient for the joint concerned depends on the wave period and hence on its height. A simple equation is also adopted here:

$$K = \frac{a + bH}{1 + cH}$$

where a, b, c are three constants depending on the joint considered, to be determined from the results of three dynamic calculations. This gives the histogram of the nominal stress ranges in the joint considered, in the following form:

$$\Delta\sigma = \frac{a + bH_o(1-\ln(N)/\ln(N_o))}{1 + cH_o(1-\ln(N)/\ln(N_o))} \alpha H_o^\beta (1-\ln(N)/\ln(N_o))^\beta$$

Since it is implicity assumed that a stress cycle is caused by the passage of a wave, $N_o$ represents the total number of stress cycles and N the number of stress cycles for which the stress range is greater than $\Delta\sigma$.

In these conditions, since $H_o$ and $N_o$ are known from the site data, five constants have to be determined from the results of three dynamic calculations concerning three wave heights and three different periods.

When dynamic analyses are not conducted (overall dynamic amplification coefficient assumed to be 1), two static calculations are sufficient. To estimate correctly the stress range in a joint during a cycle, at least six different wave positions must be examined in succession. This means that twelve static calculations are actually performed.

*This method can be used when dynamic effects are slight (the natural period of the structure is sufficiently lower than the wave periods). It applies to a histogram of wave heigths of any form.*

*In the formula $\Delta\sigma_s = \alpha H^\beta$, $\alpha$ and $\beta$ depend on the geometry of the structure and of the joint considered.*

*The coefficient K decreases with the period and tends toward 1 for high periods (Fig. 2.8).*

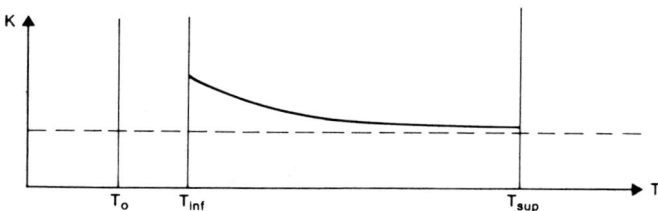

$T_o$ : natural frequency of the structure.
$T_{inf}$ : lowest wave period.
$T_{sup}$ : highest wave period.

*Fig. 2.8.*

*This approach assumes that the number of waves $N_O$ is equal to the number of stress cycles. Thus the secondary stress cycles due to the different nonlinearities and to dynamic effects are ignored.*

*For the dynamic calculation, use is made of the methods described in Section 2.2.3 (linear behavior) or in Section 2.2.4 (nonlinear behavior). Three clearly distinct waves are selected, such as the 100 year wave, the project design wave and a low wave.*

*Table 2.1.*

Mobile structures

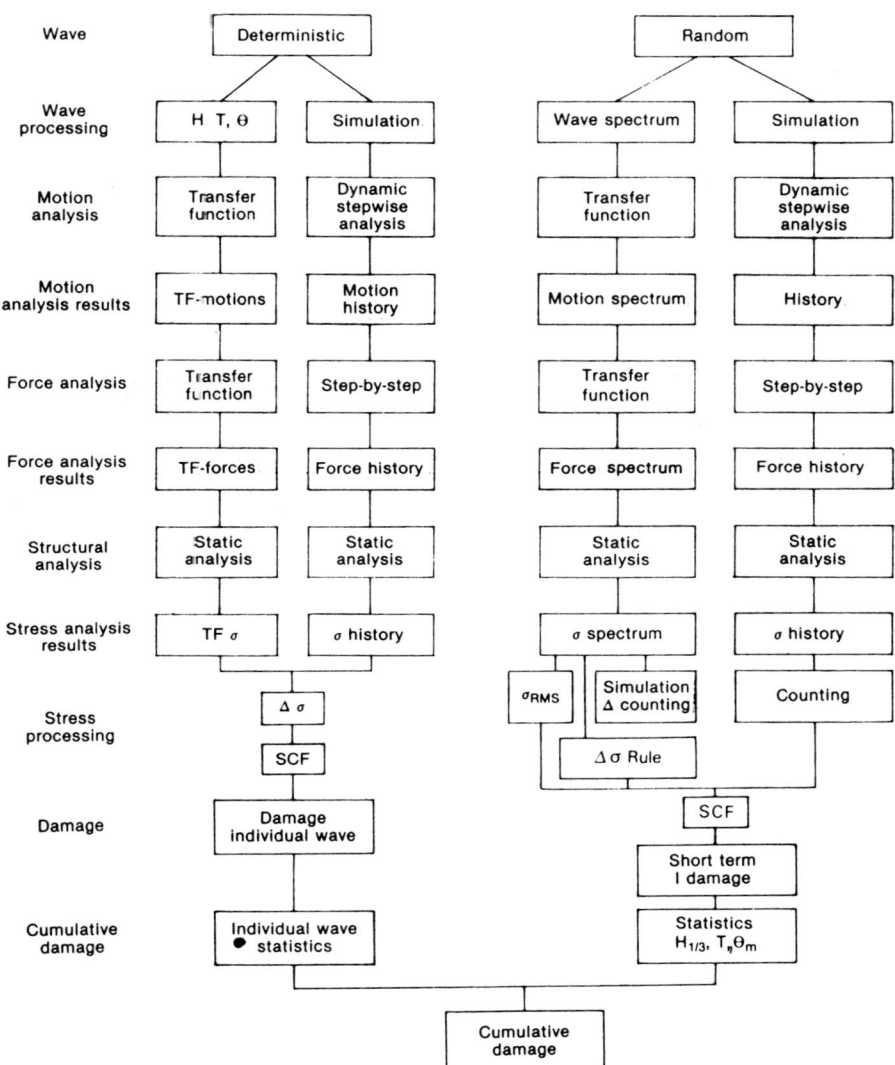

*Table 2.2.*

Fixed structures

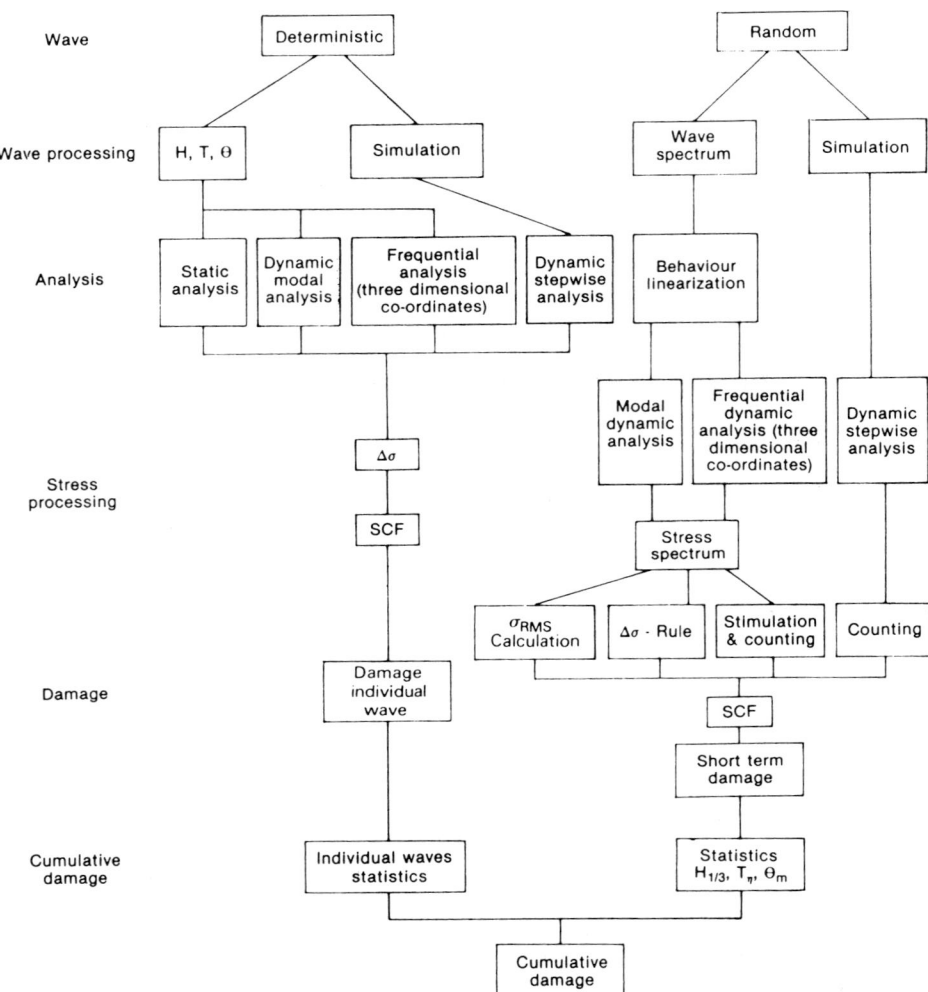

## REFERENCES

2.1   Rules for the Design, Construction and Inspection of Offshore Structures, DnV, 1977, reprinted 1979 with Appendices.

2.2   Recommended Practice for Planning, Designing and Constructing Fixed Offshore Platforms, API/RP2A, 1982.

2.3   Rules and regulations for the Construction and Classification of Offshore Platforms, Bureau Véritas, 1975.

2.4   Regulations for the Structural Design of Fixed Offshore Structures on the Norwegian Continental Shelf, NPD (unofficial translation), 1977.

2.5   Offshore installations, Guidance on Design and Construction, Proposed New Fatigue Design Rules for Steel Welded Joints in Offshore Structures, DOE, 1981.

2.6   Code of Practice for Fixed Offshore Structures, British Standards Institution, BS 6235:1982.

2.7   Wehausen, J.W. and Laitone, E.V., Surface Waves, Encyclopaedia of Physics, Vol.IX.

2.8   Borgman, L.E., Ocean wave simulation for engineering design, ASCE WW4, 1969.

2.9   Structures en mer, Dimensionnement, fabrication, comportement des structures en acier, CNEXO/CTICM, September 1976.

2.10  Miller, B.L., Wave slamming on offshore structures, National Maritime Institute, March 1980.

2.11  Hogben, N. and Lumb, H., Ocean Wave Statistics, HMSO, London, 1967.

2.12  Nordenstroem, N., Methods for predicting long-term distributions of wave loads and probability failure for ships, Part 1, Environmental conditions and short-term response, Det Norske Veritas, 1971.

2.13  Nordenstroem, N., A method to predict long-term distribution of waves and wave-induced matrix and loads on ships and other floating structures, DnV Publication No.81, 1973.

2.14  Cavanie, A., Arhan, M. and Ezraty, R., A statistical relationship between individual heights and periods of storm waves, BOSS 1976, Vol.II, p. 354.

2.15  Longuet-Higgins, M.S., On the statistical distribution of the heights of sea waves, Journal of Marine Research, Vol.II, No.3, 1952.

2.16  Houmb, O.G. and Overvik, Parametrization of wave spectra and long-term distribution of wave height and period, BOSS 1976, Vol.I, p. 144.

2.17  Daubert, A., De la Houle aux Modèles Probabilistes de la Mer, Mécanique des Fluides Appliquée, published under the direction of Michael Hug, Eyrolles, 1975.

2.18  St Denis, M., On the reduction of motion data from model tests in confused seas, Proceedings of the Symposium on the Behavior of Ships in a Seaway, Wageningen, 1957.

2.19  Pierson, W.J. and St Denis, M., On the motion of ships in confused seas, Trans. SNAME, Vol.61, 1953.

2.20  Warnsinck, W.J., Report of Committee I, International Ship Structures Congress, Delft, 1964.

2.21  Walden, H., Die Eigenschaften der Meereswellen im nordatlantischen Ozean, Statistik 10-jähriger Seegangsbeobachtungen der nordatlantischen Ozean-Wetterschiffe, Deutscher Wetterdienst-Seewerreramt, Hamburg, Einzelveröff.41, 1964.

2.22  Roll, H.U., Höhe, Länge und Steilheit der Meereswellen im Nordatlantic, Deutscher Wetterdienst-Seewetteramt, Hamburg, Einzelveröff.1, 1953.

2.23  Price, W.G. and Bishop R.E.D., Probabilistic Theory of Ship Dynamics, Chapman and Hall, London, 1974.

2.24  Maddox, N.R. and Wildenstein, A.W., A spectral fatigue analysis for offshore structures, OTC, Paper No.2261, 1975.

2.25  Marshall, P.W., Dynamic and fatigue analysis using directional spectra, OTC. Paper No.2537, 1976.

2.26  Vughts, J.H. and Kinra, R.J., Probabilistic fatigue analysis of fixed offshore structures, OTC, Paper No.2608, 1976.

2.27 Vughts, J.H., Mines, I.M., Natajara, R. and Schumm, W., Modal superposition vs direct solution techniques in the dynamic analysis of offshore structures, BOSS, 1979.

2.28 Penzien, J. and Berge, B., Three-dimensional stochastic response of offshore towers to wave forces, OTC, Pper No.2050, 1974.

2.29 Numerical methods in Offshore Engineering, Chapters 2, 3, 6, 7, 8 and 9, Edited by O.C. Zienkiewicz, R.W. Lewis and K.G. Stagg, John Wiley and Sons, 1978.

2.30 Godeau, A.J. and Deleuil, G.E., Dynamic response and fatigue analysis of fixed offshore structures, OTC, paper No.2260, 1975.

# Determination
# of the Stress Concentration Factor
# in Simple Geometry Joints

## 3.1 INTRODUCTION

The three methods routinely used to determine the stress concentration factor (SCF) are the following:

**Parametric formulas** which give the value of the SCF as a function of various geometric parameters (see Section 1.2, Part I).

**The numerical method** in which the SCF of a joint is determined by means of a numerical calculation, such as the finite elements method.

**The experimental method** in which the SCF is determined by means of measurements of deformations in a laboratory model (irrespective of scale).

## 3.2  PARAMETRIC FORMULAS

The SCF can be calculated using the parametric formulas available in the technical literature [ 3.1 to 3.11 ]. It is recommended to use the following three groups of recently published parametric formulas:

(a) Formulas of Exxon production Research (EPR) [3.6].

(b) Formulas of Lloyd's Register of SHipping (Lloyd's) [3.7].

(c) Formulas of Det Norske Veritas (DnV) [3.8].

*The parametric formulas have been developed by curve fitting the values of the SCF as a function of the geometric parameters of the joint, the SCF values being obtained either by experimental analyses, or by numerical finite element analyses. Thus the parametric formulas provide values of the SCF as a function of various geometric parameters.*

### 3.2.1  VALIDITY OF PARAMETRIC FORMULAS

The formulas are given in Annex C, the domains of application and the validity limits for each group, as announced by their authors, being given in Table 3.1. and Table 3.2. respectively.

Three types of simple load are covered separately by these formulas, namely axial load, in-plane bending, and out-of-plane bending. From the geometric standpoint, the DnV formulas deal with T joints only, while the EPR formulas covers T, Y, K, N and KT joints, and the Lloyd's formulas cover T, Y, X, K, N and KT joints.

The three groups of parametric formulas yield the same order of magnitude of the SCF, so that it is not possible to recommend one group of formulas over another. However, the Lloyd's formulas cover a wider range of geometry and loading.

Note that the use of these formulas outside the validity limits f which they have been established by their authors is liable to introdu substantial errors.

Table 3.1.

Domain of application of EPR, DnV and
Lloyd's formulas

| Load/Type of Joint | Axial load | In-plane bending | Out-of-plane bending |
|---|---|---|---|
| T | EPR DnV Lloyd's | EPR DnV Lloyd's | EPR DnV Lloyd's |
| Y | EPR Lloyd's | EPR Lloyd's | EPR Lloyd's |
| X | Lloyd's | Lloyd's | Lloyd's |
| K, N | EPR Lloyd's | EPR Lloyd's | Lloyd's |
| KT | EPR Lloyd's | Lloyd's | Lloyd's |

Table 3.2.

Validity limits of parametric formulas

| Parameters | EPR | DnV | | Lloyd's |
|---|---|---|---|---|
| | | Chord | Brace | |
| $\alpha$ | 6.67-40 | | 7    -16 | 8    -40 |
| $\beta$ | 0.3 - 0.8 | 0.225- 0.9 | 0.3 - 0.9 | 0.13- 1.0 |
| $\gamma$ | 8.33-33.3 | 10    -30 | 10    -30 | 12    -32 |
| $\tau$ | 0.2 - 0.8 | 0.4   - 1.0 | 0.47- 1.0 | 0.25- 1.0 |
| $\Theta$ | 0.   -$\pi/2$ | $\pi/2$ | $\pi/2$ | $\pi/6$-$\pi/2$ |
| $\zeta$ | 0.01- 1.0 | | | |

*The formulas offer the fastest and least expensive means to calculate the SCF, but their use is considerably limited by their lack of generality and by the impossibility of locating the exact position of the hot spot.*

*A systematic comparison was undertaken between the parametric formulas and the test results and results from finite element analyses. The following conclusions were drawn from these comparative studies [3.9]:*

(a) Deviations are possible between the SCF values
    compared to test results and finite element results
    of around +2.0 for values of geometric parameters
    which are within the validity limits of the
    parametric formulas:

$$SCF_{FORMULA} = SCF_{TEST} \pm 2.0$$

(b) If these parametric formulas are applied outside
    the validity limits, the possible margin of error
    is much greater and may reach +4.0 in some cases:

$$SCF_{FORMULA} = SCF_{TEST} \pm 4.0$$

It is obvious that these parametric formulas are
incapable of solving the case of complex geometry joints.
For these cases, due to the lack of appropriate parametric
formulas, it is necessary to resort to a number of simplifications in evaluating the value of the SCF (see
Section 3.2.3).

The EPR and Lloyd's parametric formulas for K and KT
joints do not cover overlap joints in which ζ (=g/D) is
negative. An analysis by the method of finite elements
and experimental measurements indicate that for an overlap
joint, the SCF can be evaluated fairly accurately by
taking ζ = +0.01 in the parametric formulas [3.10].

Joints with the parameter β (=d/D) equal to 1.0 are
often used in bracing, or in the riser guide grids. Due
to the conditions inherent in fabrication (welding), the
intersection of the outer surfaces of the two tubes does
not correspond to the theoretical intersection (Fig. 3.1).

Only the Lloyd's formulas consider a value of β = 1.0,
but in this case, Lloyd's recommends taking β = 0.98 to
calculate the SCF [3.7].

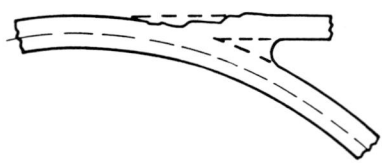

Fig. 3.1.

## 3.2.2  INFLUENCE OF GEOMETRIC PARAMETERS ON DESIGN STRESS ($\sigma_G$)

The fatigue life of a joint varies inversely with the geometric stress range, so that attempts should be made to reduce this stress. Two means are available to reduce the value of the design stress:

(a) By reducing the SCF of the joint.

(b) By reducing the nominal stress in the brace.

An examination of the influence of geometric parameters on the SCF and on the nominal stress serves to classify the possible changes in the joint geometry as follows, by decreasing order of effectiveness:

(1) Reduction in $\gamma$ and $\tau$ by increasing the chord thickness (T).

(2) Increase in $\beta$ by increasing the brace diamter (d).

(3) Increase in $\beta$ and simultaneously reduction in $\gamma$ by reducing the chord diameter (D).

The designer's attention is drawn to the technical problems raised by the use of large wall thicknesses (Chapter 3  Part I).

*The influence of various geometric parameters on the SCF has been the subject of detailed study* [3.11]. *A number of significant points in relation to a T joint are reviewed here.*

### *Influence of thickness*

*It has been observed that $\gamma$ ( = D/2T) and $\tau$ ( = t/T) are the parameters with the strongest effect on the value of the SCF. The SCF is decreased by reducing $\gamma$ and $\tau$. This may obviously result in an increase of the chord thickness T. Figures 3.2 and 3.3 show the influence of $\gamma$ and $\tau$ on the SCF for a T joint under axial load in the brace (the results are similar for a bending loading). Table 3.3 summarizes the variation in SCF as a function of T* [3.8].

*An increase in chord thickness T leads to an increase in the rigidity of the chord wall, under the action of the stress applied to the brace. This explains the importance of T as concerns the value of the SCF.*

*For a given value of $\Delta\sigma_G$, the service life decreases as thickness T increases (see Chapter 6). However, the*

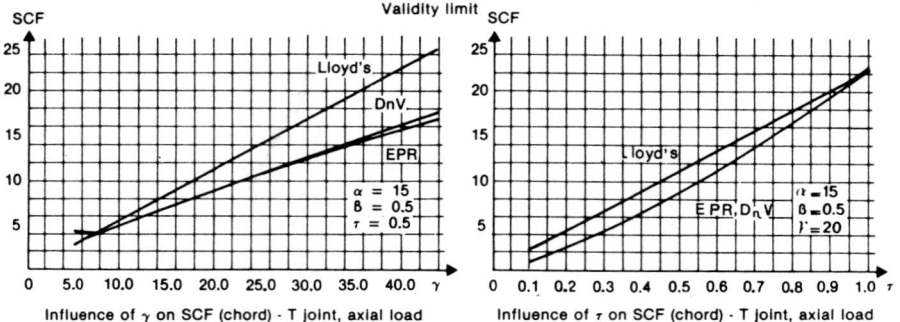

Fig. 3.2.                          Fig. 3.3.

Table 3.3.

| SCF chord | Axial load | In-plane bending | Out-of-plane bending |
|-----------|------------|------------------|----------------------|
| DnV | $(1/T)^{2.24}$ | $(1/T)^{1.43}$ | $(1/T)^{2.13}$ |
| EPR | $(1/T)^{2.141}$ | $(1/T)^{1.46}$ | $(1/T)^{1.903}$ |
| Lloyd's | $(1/T)^{2.0}$ | $(1/T)^{1.40}$ | $(1/T)^{2.00}$ |

Note: Only the value of the thickness T varies, all
      other dimensions being kept constant.

favorable influence of an increase in T on the SCF and
hence on $\Delta\sigma_G$ more than offsets this first effect,
resulting in a considerable increase in the service life
as T increases. This explains the importance of joints
with stubs.

On the other hand, large plate thicknesses raise
problems of fabrication and inspection (hardness, lamellar
tearing, anisotropy, rolling residual stresses, etc.).

### Influence of the parameter β (=d/D)

The influence of the parameter β (=d/D) is generally
weaker than that of γ or τ. An increase in β leads to a
reduction in the value of $\Delta\sigma_G$ for a given loading, either

by increasing the joint rigidity, or by decreasing the nominal stress in the brace (if d is modified). To reduce the value of $\Delta\sigma_G$, it is more effective to increase the value of d (brace diameter) than to reduce D (chord diameter).

### Joint service life

For an S-N curve with slope m, one can write that the fatigue life N of the joint varies proportionally with the following approximate expression:

$$[(1/\beta)^a \cdot (1/d)^b \cdot (1/T)^c \cdot \Delta F]^m$$

where $\Delta F$ represents the variation in a given axial or bending load. The expression does not account for a load combination.

Table 3.4 gives the coefficients a, b and c according to the type of load, as well as the relative variation in N for a variation of +10% in each of the parameters $\beta$, d, T, for m = -3.

Table 3.4.

| Load | a | b | b | Influence of $\beta$ (%) | Influence of d (%) | Influence of T[1] (%) |
|------|-----|-----|------|-----------|-----------|-----------|
| Axial load | 2.0 | 0.0 | 2.0 | +77 | +77 | +65 |
| In-plane bending | 0.5 | 1.0 | 1.45 | +15 | +53 | +40 |
| Out-of-plane bending | 0.5 | 1.0 | 2.0 | +15 | +53 | +65 |

(1) The influence of T on the S-N curve has been taken into account.

The above results from simplifications and its only purpose is to clearly highlight the parameters which exert the greatest influence on fatigue life. It does not replace the parametric formulas and is intended only as an *aid to help the designer to optimise the joint geometry in terms of fatigue strength.*

## 3.2.3 EVALUATION OF THE DESIGN STRESS RANGE IN A JOINT UNDER COMPLEX LOADS

The parametric formulas concern the evaluation of the SCF for each type of simple loading. The design stress range for a complex loading (combined loads) can be evaluated by the following equation:

$$\Delta\sigma_G = SCF_{Ax} \cdot \Delta\sigma_{Ax} + SCF_{Fy} \cdot \Delta\sigma_{Fy} + SCF_{Fz} \cdot \Delta\sigma_{Fz}$$

where

$SCF_{Ax}$ = stress concentration factor for axial loading,

$SCF_{Fy}$ = stress concentration factor for an in-plane bending load in the joint,

$SCF_{Fz}$ = stress concentration factor for an out-of-plane bending load in the joint,

$\Delta\sigma_{Ax}$ = axial nominal stress range in the brace,

$\Delta\sigma_{Fy}$ = nominal joint in-plane bending stress range in the brace,

$\Delta\sigma_{Fz}$ = nominal joint out-fo-plane bending stress range in the brace.

The influence that an axial load in the chord may exert on the SCF is generally slight, and is therefore ignored in the formula. However, for joints in which the nominal axial stress in the chord is of the same order of magnitude as the nominal stress in the brace and where the hot spot on the chord occurs near the crown point, the value of the nominal axial stress range in the chord must be added to the geometric stress range in the chord calculated by the proposed method.

*The possibilities of calculating the exact SCF under combined loading are very limited. Strictly speaking, the SCF values for simple loadings can be combined linearly only if the principal stress is a maximum ($\sigma_G$) at the same hot spot and in the same direction for each of the simple loadings.*

*The major drawback of parametric formulas is they provide only the value of the geometric stress at the hot spot, for a given simple loading. The stress distribution*

*along the weld remains unknown. Experimental analyses and numerical calculations by finite elements show that the hot spot lies close to the saddle point for an axial or out-of-plane bending load, and near the crown point for an in-plane bending load. Similarly, the stress is approximately perpendicular to the weld in the neighborhood of the crown point and the saddle point. Finally, a zone exists along the weld and on either side of the hot spot, in which the value of $\sigma_G$ varies little from value at the hot spot [3.8]. Since the extent of this zone depends on the joint geometry and the loading (Fig. 3.4), no simple, general method is available to account for the stress distribution along the weld. This is why the method proposed assumes that $\sigma_G$ has the same value, under simple loading, at every point of the brace-chord junction.*

*For this reason the formula is largely conservative. A more accurate value of $\sigma_G$ can be obtained by using the approach proposed in Section 3.3.*

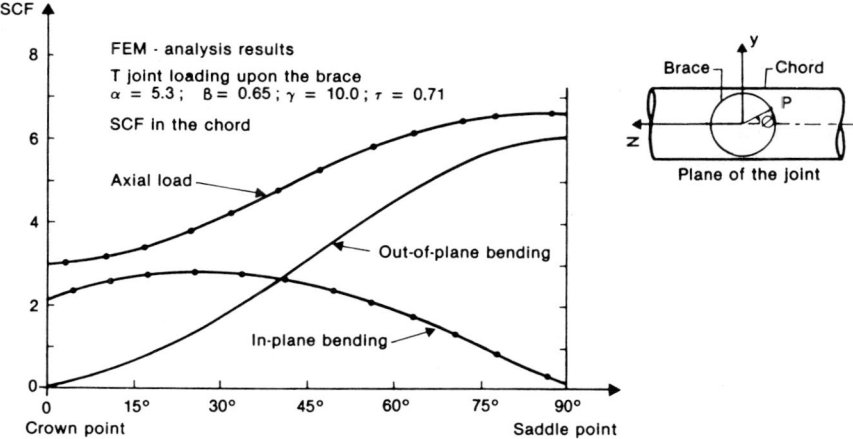

*Fig. 3.4.*

## 3.3  NUMERICAL METHOD

### 3.3.1  GENERAL

Among all the numerical analysis methods available for structural calculations, the finite element method (FEM) is the most widely used, thanks to its versatility and reliability.  However, any calculation by FEM requires some experience on the part of the user.

The evaluation of the SCF in simple geometry joints raises a number of delicate questions, which are discussed below with indications being provided on the choices to be made.

*A number of analytical approaches have been developed to study the stress distribution in very simple tubular joints (such as T joints) with specific loads [3.13, 3.13, 3.14].  These analytical approaches, which results from very fragmentary investigations, are far from offering the general validity and flexibility of use as provided by the finite element method.*

### 3.3.2  PROCEDURE FOR THE NUMERICAL ANALYSIS OF THE STRESS CONCENTRATION IN A TUBULAR JOINT

In actual fact, tubular joints are complex structural elements.  From a modelling standpoint, an attempt is made to take account of this reality, so as to locate the stress concentrations as accurately as possible.  The finite element method is excellent for taking into account of specific geometric details such as shells and stiffeners, which may play a very important role in reducing the stresses, but may also make it more difficult to locate the hot spot.

#### A.  General approach to calculate the stress concentration

For more or less complex tubular joints, a preliminary calculation by classic analytical methods are required to determine the boundary conditions in terms of forces and displacements.

After having determined these boundary conditions, the joint to be analyzed is isolated from the structure and modelled in finite elements. The boundary conditions previously calculated are applied to the ends of the joint thus isolated from the rest of the structure.

Analysis using finite elements helps to obtain the maximum principal stresses in the tube-wall surfaces. The maximum value of the maximum principal stress, called the maximum geometric stress, is representative of a given geometry and loading.

## B. Approach in case of poorly known boundary conditions

Insofar as the boundary conditions are poorly known, as in the preliminary design stage, special attention must be paid to the length of the chord section adopted in T, K and KT joints.

To determine the stress concentration at the brace-chord junction, the tube lengths beyond the junction are determined to ensure that the calculated SCF is unaffected by the boundary conditions, in the sense of the application of the Saint Venant principle. It is recommended to use the following minimum lengths (Fig. 3.5):

(a) For the main chord, lengths of at least 1.5 times the chord diameter, starting from the cross-sections passing through the crown points at the brace(s).

(b) For the brace, a length of at least twice its diameter.

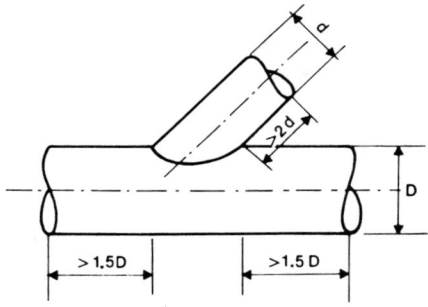

Fig. 3.5.

*The tubular joint under consideration forms part of the structure. Before "detaching" this tubular joint from the rest of the structure for a detailed analysis of the stress distribution, it is necessary to calculate the loads acting on the different ends of the tubular joint, for different possible wave action situations.*

### 3.3.3. COMPUTERS PROGRAMS

It is difficult to recommend the use of a specific computation software. However, considering the remarks expressed in the comments, it is worth noting that the use of a software to solve problems of stress concentration determination is considerably facilitated by the "utilities" available for joint meshing and for providing a fast, clear analysis of the desired results, namely :

(a) Location of the hot spot.

(b) Magnitude of the stress concentration.

*In practice, several finite element programs are available to solve problems of shells in linear elastic media. These programs differ mainly in the following:*

*(a) The global geometry in the finite elements used (flat elements, curved elements, three-dimensional elements). As a rule, the finite elements of the "thin shell" type yield excellent results. Three-dimensional finite elements do not offer better results.*

*(b) Formulations concerning the displacement, deformation and stress fields of the elements. In this respect, the elimination of the torsional rigidity about the normal to the mean surface exerts a relatively strong influence in certain cases.*

*(c) The technique adopted to solve the problem of singularity at the intersection of two shells [3.15, 3.16].*

*(d) The general resolution method adopted; minimum potential energy, minimum complementary energy and the different hybrid methods [3.17].*

### 3.3.4 MESHING OF TUBULAR JOINTS

From the purely geometric standpoint, two types of finite elements are normally used: flat elements for which the membrane and bending effects are superimposed [ 3.17 ] , and single or double curvature elements (superparametric and subparametric elements [ 3.18], elements by Visser [3.19], etc.

The use of any of these types of element leads to comparable results, provided care is taken in the mesh configuration. The use of the flat element certainly requires a relatively fine meshing for an accurate matching of the curved surfaces of the tubes making up the tubular joint.

In selecting the mesh dimension, it is recommended to observe the following rules [3.20]:

(a) For the finite elements in the neighborhood of the intersection of the tubes:
- The dimension of the side of the element perpendicular to the intersection curve of the two tubes (denoted m or e on Fig. 3.6) must be such that the length a to the centre of gravity of the element is located no further than 0.4T from the "imaginary" weld toe projected onto the mean surfaces. The lower of the two values m and e is then used.
- The dimension b of the element whose side lies on the intersection curve of the T or Y joint must be less than 1/24 of the length of the intersection curve.

(b) With respect to the dimension of the mesh sides far from the intersection zone:
- The maximum dimension of the sides of the finite element must not exceed the half-radius of the tube on which it is located.
- The passage from the smallest finite elements to the largest must take place gradually.

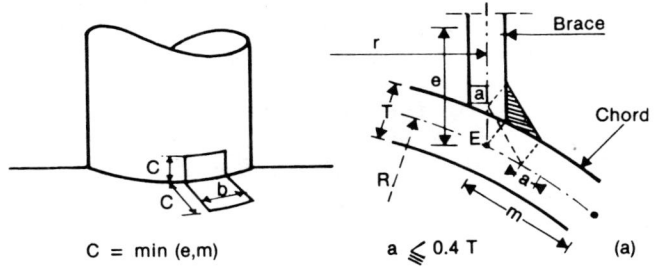

$$C = \min (e,m) \qquad\qquad a \leqslant 0.4\ T \qquad (a)$$

Fig. 3.6.

**Specific case of joints for which  $\beta = d/D = 1$**

Figure 3.7 shows details of the intersection of two tubes. To mesh the intersection of this type of joint, the following method is proposed [3.23, 3.24]:

(a) Determine point a, the intersection of the inner wall of the brace and the outer wall of the chord.

(b) Project point a onto the mean surface of the brace, i.e. a'; the length is determined in this way.

(c) Determine point b on the mean surface of the chord at mid-height (1/2).

The meshing in the neighborhood of the intersection must follow the broken line c b a'. Hence the mesh size must not exceed the length cb or ba'.

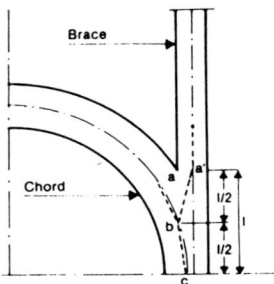

Fig. 3.7.

*Meshing is a delicate matter in any calculation by the finite element method. Although computation costs increase approximately with the square of the number of elements, the meshing must not be too large, because the accuracy of the results depends directly on the element size.*

*The fineness of the mesh also depends to a great extent on the formulation employed (displacement fields, stress fields, or hybrid), to guarantee the best continuity of the displacements, slopes and curvatures and the element boundaries.*

*For tubular joints, the smallest meshes lie in the neighborhood of the tube intersection. Since it is often difficult to predict the most highly stressed point, especially for complex loads, meshes of approximately equal size are created about the intersection if possible.*

*At the intersection of two tubes, the real thickness and presence of the weld create a zone whose behavior is strongly three-dimensional (Fig. 3.8). The calculation of the stress concentration at the intersection of two tubes remains associated with a conventional definition (see Section 3.3.5).*

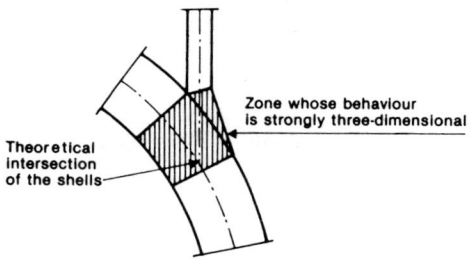

*Fig. 3.8.*

The centres of gravity of the finite elements used in the neighborhood of the intersection should lie outside the zone shown in Fig. 3.8.

If a variation exists in the tube wall thickness (case of cast joints), it is better to use shell finite elements, which allow one to associate a specific thickness with each node (Fig. 3.9).

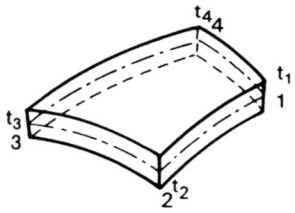

*Fig. 3.9.*

The recommendations concerning the fineness of the mesh and the interpretation of the numerical results obtained by finite element analysis result from comparisons between the numerical results and the experimental results, given in the Ref. [3.21, 3.22].

## 3.3.5 INTERPRETATION OF RESULTS

The stress concentration is estimated from the value of the maximum principal stress. Computer programs must therefore give the values of the principal stresses on the inner and outer surfaces of the tubes.

The principal stresses are preferably calculated at a point on the surface corresponding to the centre of gravity of the finite element, so that only one mean value of the stress tensor is obtained for each finite element.

The value of $\sigma_G$ (see Section 1.2) is evaluated as follows:

(a) For the chord: the value of the maximum principal stress is extrapolated or interpolated to the weld toe (see Fig. 3.10).

(b) For the brace: the value of the maximum main stress is also extrapolated or interpolated to the weld toe (see Fig. 3.11). According to a comparative study of finite element analyses and experimental measurements, the value thus calculated is low.

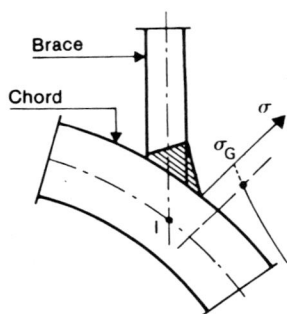

Fig. 3.10.

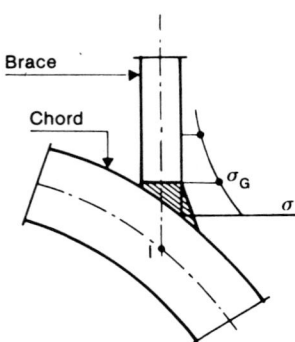

Fig. 3.11.

*If the computer program gives the values of the principal stresses at the Gauss points, it is always possible to relate these values to the centre of gravity of the finite element. As a rule, computer programs give the values at the centre of gravity directly.*

*It has been observed that the stress concentration values at the intersection of the mean surfaces of the tubes (point I in Fig. 3.12) calculated by the finite element method are greater than those obtained at the weld toes (chord side and brace side), according to the experimental methodology discussed in Section 3.4. To correct the value obtained by finite element method, it is proposed to estimate this value at a point other than the intersection of the mean surfaces.*

*All the parametric formulas developed by EPR and DnV [3.6, 3.8] are based on the results of an analyses with finite elements. The authors of these formulas have estimated the values of the SCF at the hot spot points shown in Fig. 3.13 for EPR and Fig. 3.12 for DnV.*

*Note that the values given by the extrapolations considered, either by DnV or by EPR, do not always correspond to the experimental values on actual joints. Significant discrepancies have been observed, which appear to depend on the geometric parameters. When establishing their parametric formulas, DnV subsequently reduced by 25% the value of the brace SCF given by the finite element analyses results for this reason [3.8].*

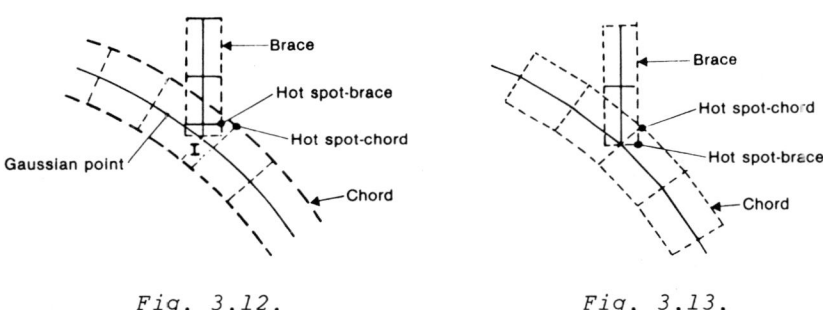

Fig. 3.12.                              Fig. 3.13.

## 3.3.6  EVALUATION OF THE DESIGN STRESS IN A JOINT UNDER COMPLEX LOADS

In the fatigue analysis of the joint, the three types of load (axial load, in-plane bending, out-of-plane bending) in the chord and in the brace (see Section 1.2) must be considered in combination in calculating the design stress by the finite element method.

*The FEM is ideal for determining the SCF in a joint under complex loading. For each simple load case, the analysis furnishes complete data on the values of the stress and deformation at each mesh. Load combinations are easily processed by superposition, using the computer.*

## 3.4  EXPERIMENTAL METHOD

*The experimental analysis of a joint model should be entrusted to a specialised laboratory with the experience necessary for this type of test. To obtain valid results, it is essential to exercise special care in selecting and checking the boundary conditions.*

*At present, these experimental analyses are extremely expensive, and this limits their use to research activities.*

### 3.4.1  STRAIN GAUGE MEASUREMENTS

The experimental method generally followed consists in applying simple loads to the joint model to be analysed. For each of these loads, the position and value of the stress at the hot spot is determined, from the data supplied by electrical resistance strain gauges [3.25 to 3.28].

The model analysed may be of steel or of another material (acrylic, plycarbonate) exhibiting mechanical properties that allow the measurement of steep strain gradients for a linear elastic state.

The stresses around the weld toe are determined by linear extrapolation of the values of the maximum principal stress $\sigma_G$ on the outer surface of each tube.

To accurately determine the position of the hot spot, it is necessary to analyse several areas around the weld, on the brace side and on the chord side.

Figure 3.14 shows in detail the extrapolation required to obtain the value of the geometric stress $\sigma_G$. The figure shows how to construct the extrapolation line passing through points $A_i$ and $B_i$, for the saddle and crown points.

For certain geometries and certain loads, the hot spot is not necessarily located on lines 1, 2, 3 and 4 shown in the figure. Based on the same principles stated above, it is possible to analyse along the line of intersection of any plane passing through OO' and the outer walls of the tubes.

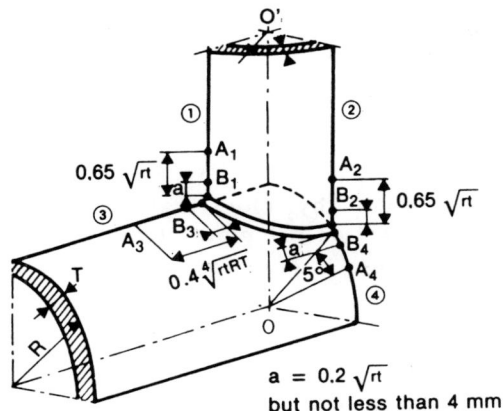

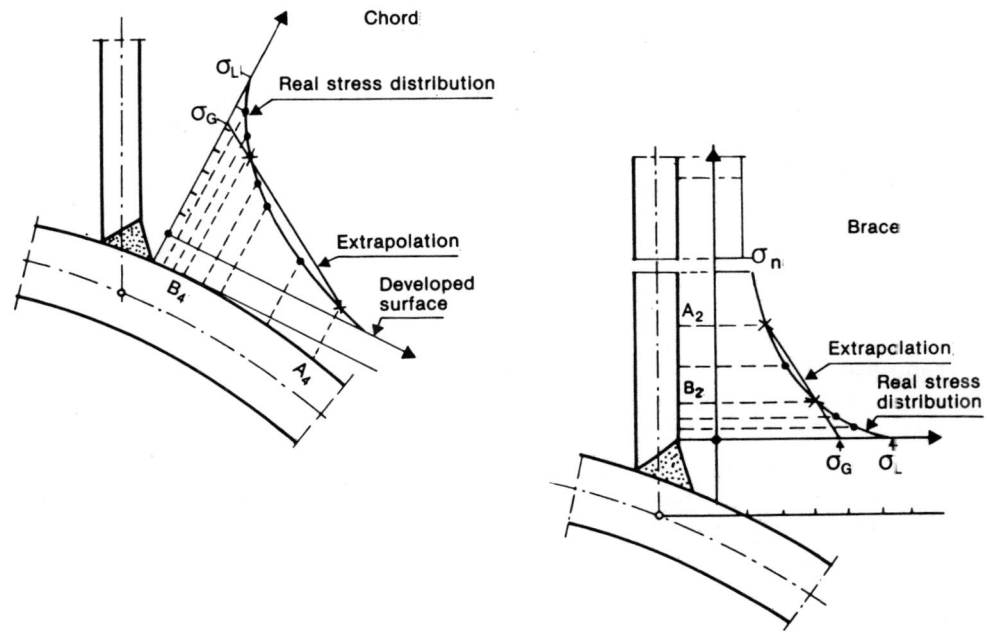

Fig. 3.14.

The stress concentration factor is written:

$$SCF = \frac{\sigma_G}{\sigma_n}$$

where $\sigma_n$ is the nominal stress (see Section 1.2).

The determination of the stress concentration factor (SCF) at the hot spot from measurements given by electrical resistance strain gauges requires the knowledge of the two principal strains $\varepsilon_1$ and $\varepsilon_2$. The value of the maximum principal stress is given by the formula:

$$\sigma_1 = \frac{E}{(1-\nu^2)} (\sigma_1 + \nu\sigma_2)$$

where E is Young's modulus of the material and $\nu$ the Poisson ratio. If $\varepsilon_1$ and $\varepsilon_2$ are not measured at the same points, the values corresponding to points $A_i$ and $B_i$ can be determined by interpolating the values of $\varepsilon_1$ and $\varepsilon_2$.

If no data is available on the value of $\varepsilon_2$, the following can be used as a close approximation:

$$SCF = 1.15 \ SNCF$$

where SNCF is the strain concentration factor at the hot spot. It is determined in a similar manner to the SCF, by linear extrapolation of the values of $\varepsilon_1$ at points $A_i$ and $B_i$. The deformation concentration factor is written:

$$SNCF = \frac{\varepsilon_G}{\varepsilon_n}$$

where $\varepsilon_n$ is the value of the nominal strain and $\varepsilon_G$ is the value of $\varepsilon_1$ extrapolated to the hot spot.

*During a static load test, several load levels are applied in steps, and the strain distribution is observed at each step. As a rule, for steel models, several successive loadings are carried out (loading and unloading cycles), in order to detect any "shake down" of the structure. It is important that the measurements for the determination of the SCF correspond to an elastic behavior of the model at the stress concentration zone. Fig. 3.15 shows a typical loading history of a static load test [3.26].*

*The method for determining the SCF described in the recommendations results from decisions taken by the ECSC community programme working groups. The method gives a conventional definition of the geometric stress. The extrapolation method, which was developed for the case in which the hot spot lies at the crown point or at the saddle point of the joint, can be extended to the case in which the hot spot lies between these two positions.*

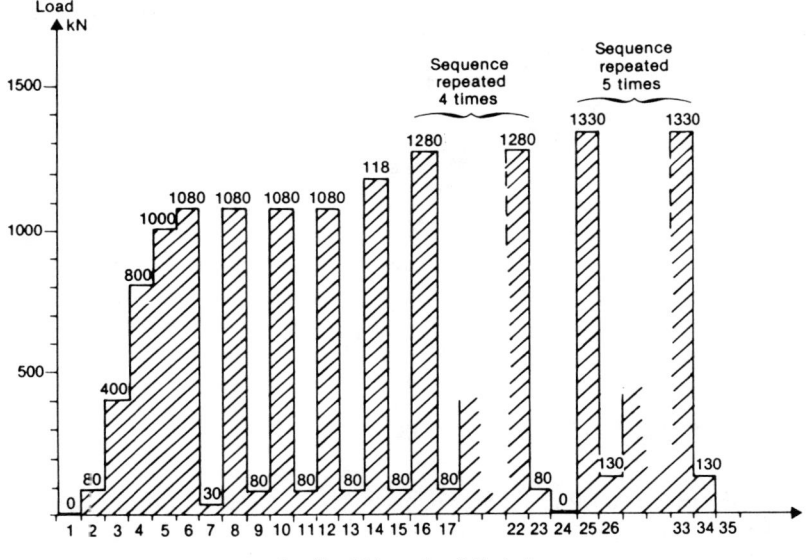

Loading history of a static test

*Fig. 3.15.*

*If the position of the hot spot and the directions of the principal stresses are known in advance, it suffices to place the strain gauges on a line culminating in the hot spot. If not, it is possible to explore the values of the principal strains on the brace and chord lines leading to the saddle and crown points, and if necessary, at a number of intermediate points of the chord-brace intersection. The difference between the highest stress obtained at the saddle point or the crown point and the stress value at the hot spot located at an intermediate point of the weld is generally small.*

*In connection with measurements on small steel models, it must be recalled that the measurement of the SCF may be erroneous due to the disproportion of scale in comparison with real conditions between the weld and the joint geometry. Very often, welds do not exist on acrylic models.  In this case, the hot spot generally lies at the intersection, as shown in Fig. 3.16.  The Lloyd's parametric formulas have been developed from tests on strain-gauged acrylic models [3.7, 3.25].*

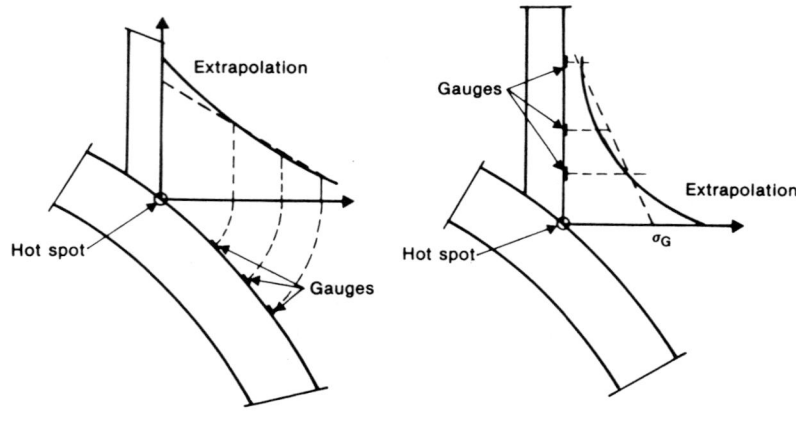

*Fig. 3.16.*

## 3.4.2 PHOTOELASTIC ANALYSIS

This method, which is often used to analyse tubular joints, essentially consists of "freezing the stresses" in a three-dimensional model [3.28 to 3.34]. The model is prepared in a mould which perfectly reproduces the connection profiles between the chord and brace walls, and if applicable, the weld profile.  The model is built of a "photoelastic" material (for example, epoxy resin) and heated to the "freezing" temperature defined accurately for each material (about 100 to 150°C).

Once the loading is applied, the model is cooled slowly (2°C/h), under load, to ambient temperature.  After the load is removed, the stresses, deformations and accompanying photoelastic effects (birefringence) induced by the loading, remain "frozen" in the cooled model.

Photoelasticity techniques [3.30] serve to determine the values and the directions of the principal stresses in thin slices cut out of the model along preferential planes.  This technique, which is destructive, is not the only one available.

The values of the design stress $\sigma_G$ and the SCF are obtained by the extrapolation method (see Section 3.4.1).

*The advantage of this method resides in the possibility of perfectly representing the connection profiles between the brace and the chord. The method yields excellent results, but demands considerable experience on the part of the operator.*

*The method can be used not only to determine the SCF in a joint, but also to examine the influence of the weld profile (Fig. 3.17).*

*As a rule, neither the individual passes nor the local effects (undercuts, variation in profile along the weld, etc.) are modelled. The figures show the influence of the weld toe profile on the local stresses at the weld toe [3.29, 3.31]. For fatigue calculations, the value of the geometric stress (see Chapter 1) is used rather than the value of this local stress.*

*The main drawback of the slicing method is the destruction of the model and the inability to analyse other loadings on the same model. Hence a faily good idea of the direction of the principal stress planes must be obtained before slicing.*

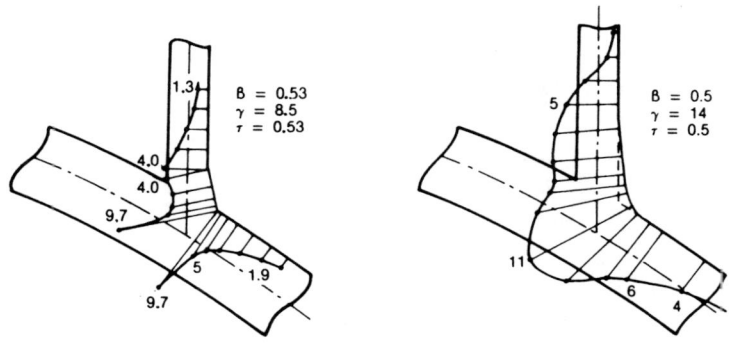

*Fig. 3.17.*

## 3.4.3 EVALUATION OF THE DESIGN STRESS IN A JOINT UNDER COMPLEX LOADS

The application of complex loads often raises technical problems in the laboratory. The problem can be solved by considering each type of simple load separately. If the elastic behavior of the structure is guaranteed, cases of combined loads can be dealt with by superimposing the measurements.

*Due to the limitations prevailing at the technical level (equipment, boundary conditions) and also from the economic standpoint, the treatment of complex loading cases on the basis of strain gauge measurements is rarely feasible. The technique of superimposing simple loads is therefore often adopted, but since only partial data are available for each simple load case, the method is not always capable of solving complex load cases (for example, if the hot spot for the complex load lies neither at the crown point nor at the saddle point).*

## REFERENCES

3.1    Beale, L.A. and Toprac, A.A., Analysis of in-plane T, Y and K welded tubular connections, Welding Research Council Bulletin No.125, May 1968.

3.2    Reber Jr., J.B., Ultimate strength design of tubular joints, OTC, Paper No.1664, 1972.

3.3    Visser, W., On the structural design of tubular joints, OTC, Paper No.2117, 1974.

3.4    Marshall, P.W. and Graff, W.F., Limit state design of tubular connections, BOSS 1976, Trondheim, Norway.

3.5    Kinra, R.K. and Marshall, P.W., Fatigue analysis of the Cognac platform, OTC, Paper No.3378, 1979.

3.6    Potvin, A.B., Kuang, J.G., Leick, R.D. and Kahlich, J.L., Stress concentrations in tubular joints, SPE Journal, August 1977.

3.7(a)  Wordsworth, A.C. and Smedley, E.P., Stress concentrations at unstiffened tubular joints, European Offshore Steel Research Seminar, Cambridge, November 1978.

3.7(b)  Wordsworth, A.C., Stress concentrations at K, KT tubular joints, ICE Conference, Fatigue in Offshore Structural Steel, Paper No.27, London, February 1981.

3.8    Teyler, R., Gibstein, M.B., Bjornstad, H. and Haugan, G., Parametrical stress analysis of T-joints, DnV Report No.77-523, November 1977.

3.9    Ryan, I., Comparaisons statistiques entre les CCC expérimentaux et les CCC selon les diverses formules paramétriques, CTICM Report No.10.002-5, May 1982.

3.10   Moe, E.T. and Gibstein, M.B., Stress analysis and fatigue failure of K-joints with overlapping braces, DnV Technical Report No.80-1172, December 1980.

3.11   Ryan, I., Comparaisons des diverses formules paramétriques de coefficients de concentration de contraintes, CTICM Report No.10.002-2, April 1981.

3.12    Wichman, A.R., Hopper, A.G. and Mershow, J.L., Local stresses in spherical and cylindrical shells due to external loadings (Bijlaard method), Welding Research Council Bulletin No.107, August 1965, reprinted June 1977.

3.13    Dundrova, V., Stresses at intersection of tubes, Cross and T-joints, SFRL Report, University of Texas, 1966.

3.14    Scordelis, A.C. and Bouwkamp, J.G., Analytical study of tubular T-joints, ASCE Journal, Structural Division, January 1970.

3.15    Greste, O., Finite element analysis of tubular K-joints, Structural Engineering Laboratory, University of California, Berkeley, Report No.UCESM 70-11, June 1970.

3.16    Zienkiewiecz, O.C., The Finite Element Method in Engineering Sciences, McGraw-Hill, London, 3rd Edition, 1977.

3.17    Gallagher, R.H., Introduction aux Eléments Finis, Editions Pluralis (french version), 1976.

3.18    Zienkiewicz, O.C., Irons, B.M., Ergatoudis, J., Ahmad, S. and Scott, F.C., Iso-parametric and associated element families for two- and three-dimensional analysis, in Finite Element Methods in Stress Analysis by I. Holland and K. Bell, Tapir, Norway, 1970.

3.19    Visser, W., The application of a curved mixed-type shell element, Proceedings Symp. High-speed Computing of Elastic Structures, Liège University, Belgium, 1970.

3.20    Recho, N. and Brozzetti, J., Concentration de contraintes dans les piquages de tubes due à des sollicitations axiales, CTICM Report No.10-002-6, May 1982.

3.21    United Kingdom Offshore Steels Research Project, Final Report to ECSC, Agreement No.7210 KB/8/801, Department of Energy, United Kingdom, April 1981.

3.22    Gérald, J., Interprétation des essais sur les noeuds IRSID, Laboratoire de Mécanique des Solides, ANMT Technical Report No.15, 1981.

3.23    Dijkstra, O.D., Visser, W., Janssen, G.T.M., Comparison of strain distributions in three X-joints determined by strain gauge measurements and finite element calculations, ECSC/IRSID International Conference, Steel in Marine Structures, Paris, October 1981.

3.24 Mezière, Y., Etude d'un joint en K avec recouvrement, ANMT Technical Report No.19, Laboratoire de Mécanique des Solides, September 1982.

3.25 Wordsworth, A.C., Experimental determination of stresses at unstiffened tubular T and X-joints, Joint Australian Welding and Testing Conference, Perth, October 1977.

3.26 Lieurade, H.P. and Gérald, J., Résultats des essais statiques de dix noeuds en X en vraie grandeur, ECSC/IRSID International Conference, Steel in Marine Structures, Paris, October 1981.

3.27 De back, J., Vaessen, G. et al, Fatigue and corrosion fatigue, Behavior of offshore steel structures (Dutch programme), Final Report to ECSC, Agreement No.7210 KB/6/602, April 1981.

3.28 United Kingdom Offshore Steels Research Project, Final Report to ECSC, Agreement No.7210 KB/8/801 for period 1 June 1977 to 31 May 1979, Vol.1 UKOSRP, Department of Energy, United Kingdom, April 1981 (British programme).

3.29 Fessler, H. and Stanley, P., Photoelastic analysis of tubular T-joints, UKOSRP Interim Technical Report 2/02, Department of Energy, United Kingdom, August 1977.

3.30 Avril, J., Encyclopédie Vishay d'Analyse de Contraintes, Vishay Micromesures.

3.31 Robert, A., Bourdon, C. and Mézière, Y., Analyse photoélastique et numérique de la concentration de contraintes dans les noeuds tubulaires, ECSC/IRSID International Conférence, Steel in Marine Structures, Paris, October 1981.

3.32 Bourdon, C., Etude photoélastique sur trois modèles de noeuds en croix de structures tubulaires soudées, STCAN, Report No.PV 3301 MSN, September 1980.

3.33 Holliday, G.H. and Graff, W.J., Three-dimensional photoelastic analysis of welded T-connections, OTC, paper No.1441, 1971.

3.34 Camponuovo, G.F. and Mondina, A., Photoelastic analysis of welded Y-joints for offshore structures, ECSC/IRSID International Conférence, Steel in Marine Structures, Paris, October 1981.

# CHAPTER 4

# Determination
# of the Stress Concentration Factor
# in Joints of Complex Geometry

In this Chapter, we shall deal with the problem of determining the SCF in three types of joints of complex geometry, routinely used in the construction of offshore structures:

(a) K and KT joints with overlap (Section 4.1).

(b) Joints with several braces (Section 4.2).

(c) Stiffened joints (Section 4.3).

Given the complexity of these joints from the geometric standpoint, no parametric formulas exist which directly give the value of the SCF. With few exceptions, it seems highly improbable that parametric formulas will be developed, even in the future. The analysis of these complex-geometry joints primarily requires numerical calculations by finite elements.

The following discussion information compiled during specific analyses of joints with a given geometry. This information is mainly intended to provide some indications about the influence of a number of parameters and specific conditions.

*The parametric formulas published in the technical literature cover simple tubular joint geometries and loading modes. Very few results are available on the determination of stress concentrations in joints with complex geometry, whether these results have been obtained numerically (finite element method) or experimentally.*

## 4.1  K AND KT JOINTS WITH OVERLAP

K and KT joints with overlap are often used to:

(a) Meet the geometric requirements imposed on the joint (other braces, etc.).

(b) Exploit the fact that the static strength of the joint with overlap is greater than that of an equivalent joint without overlap (see Chapter 2).

(c) Reduce the eccentricity of the loads applied in the braces.

(d) Reduce the stress concentrations in the joint.

In fact, very few experimental and numerical analyses have been conducted on stress concentrations in joints with overlap [4.1 to 4.5]. These analyses have not confirmed that a joint with overlap offers the advantage of a lower stress concentration than the equivalent joint without overlap [4.2, 4.4].

*Figure 4.1 shows a K joint and a KT joint with overlap. The numerical and experimental analyses of these joints indicate that the value of the SCF may be higher than 6.0, but does not exceed 4.0 in most cases [4.2].*

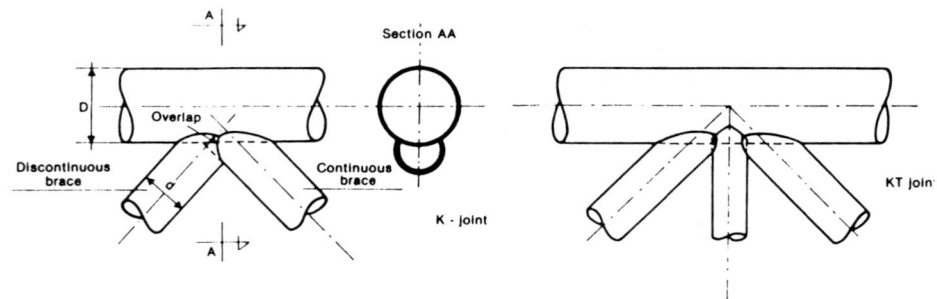

*Fig. 4.1.*

## 4.2 JOINTS WITH SEVERAL BRACES

The only way to evaluate the stress concentration in this category of joints is by numerical analysis by finite element or by an experimental method, particularly since in most cases, the presence of several braces on the same chord necessarily leads to overlaps for certain braces.

**Joints with two orthogonal braces under axial loads**

A number of studies have dealt with specific brace geometries, notably joints with axially loaded orthogonal braces [4.6, 4.7, 4.8]. These studies have revealed the influence of a number of parameters defined in Figs. 4.2 and 4.3.

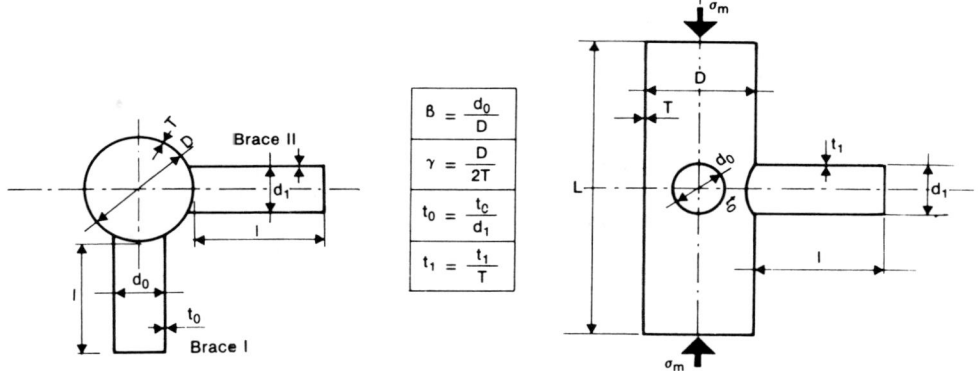

Fig. 4.2.

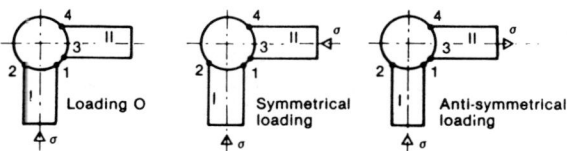

Fig. 4.3.

The conclusions summarised and presented below, which are valid for the geometry and the specific loading mode analyzed, cover the following validity range:

$$0.205 \leqq \beta \leqq 0.457$$
$$13.13 \leqq \gamma \leqq 42.00$$
$$0.365 \leqq \tau_o \leqq 0.8$$
$$0.5 \leqq \tau_1 \leqq 0.8$$
$$0.672 \leqq \rho \leqq 1.5$$

For loading 0, the influence of the rigidity provided by brace II on the value of the SCF at the intersection of brace I and the chord is negligible. For this loading, the hot spot lies simultaneously at points 1 and 2, showing that the stiffening of the chord wall provided by the presence of unloaded brace II has a negligible effect.

For symmetrical loading, depending on the relative dimensions of the braces, the hot spot lies at point 1 or point 2, or in the neighborhood of these points (in the specific case of a symmetrical geometry, the hot spot lies at points 1 and 3).

The effect due to loading of the chord is negligible in comparison with the SCF at the hot spot. However, this load has the effect of increasing the stress concentration in the brace, which is generally lower than that existing on the outer surface of the chord.

The influence of the ratio of the brace diameters $\rho = d_o/d_1$ is greater for antisymmetrical loading than for symmetrical loading (note that this effect is negligible if only one brace is loaded).

The example illustrated in the comments shows that the fatigue calculation based on the separate analysis of the SCF in each of the braces (ignoring the existence of the other) does not necessarily guarantee safety in comparison with a fatigue calculation based on a global determination of the SCF for the geometric configuration and the loading of the joint analyzed.

*The conclusions expressed in the recommendations concern the geometric and loading configuration illustrated in Figs. 4.1 and 4.2 of the recommendations, were derived from the results of numerical studies [4.6] and a number of experimental results [4.7, 4.8]. For this specific geometric and loading configuration, and due to the very fact that the numerical calculations are based on the assumption of linear elastic behavior (acceptable for calculating the SCF), the results can be presented in the form of an alignment graph [4.6].*

*Figure 4.4 is a specific graph corresponding to the parameters given in the inset. This presentation serves* to determine the SCF graphically at points 1, 2, 3 and 4, for different stress ratios R.

R is the ratio of the nominal stress of brace II to the nominal stress of brace I. This ratio varies between 1 (for symmetrical loading) and -1 (for antisymmetrical loading).

This presentation helps to visualise directly the error committed when the SCF is calculated for each brace while ignoring the existence of the other, instead of calculating the SCF by taking account of both loaded braces.

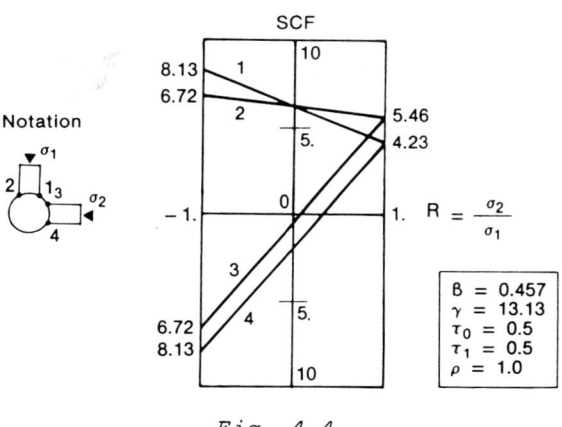

Fig. 4.4.

Let us consider the example illustrated by Fig. 4.4:

(a) Only one loaded brace
    $R = 0$ :    SCF = 6.

(b) Both braces loaded symmetrically,
    $R = 1$ :    SCF = 5.46.

(c) Both braces loaded antisymmetrically,
    $R = -1$ :    SCF = 8.13.

For a given geometric configuration, Fig. 4.2 serves to determine the SCF for any loading case. If we consider the case of $R = -0.5$: the SCF $\simeq$ 7, this value being greater than 6.00 (for a single loaded brace).

Reference [ 4.6] provides other charts valid for other geometric ratios.

## 4.3  STIFFENED JOINTS

### 4.3.1  INTRODUCTION

The stress concentrations in unstiffened joints are sometimes so high that very thick tubes are needed to guarantee the requisite fatigue life. Considering the problems associated with high thickness (see Chapter 3, Part I), many engineering firms have opted for the use of stiffeners, either to reduce the tube thickness (of the chord especially) in the stiffened joint for a given $\Delta\sigma_G$, or to reduce $\Delta\sigma_G$ by about a factor of three for the same tube thickness with the most interesting consequence of a substantial increase in fatigue life.

Two types of stiffening normally exist, either by annular stiffeners (external or internal), or by longitudinal stiffeners (see Fig. 4.5). The rest of this Section discusses joints stiffened by internal annular stiffeners only.

The effectiveness of stiffening depends on the following:

(a) The type of loading and type of stiffening, as well as the position of the stiffeners.

(b) The inertia of the stiffeners and their number.

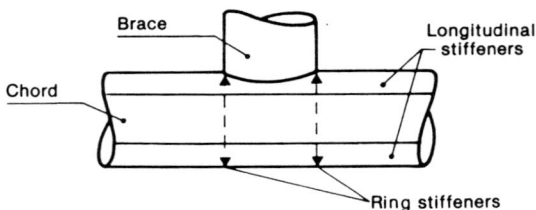

Fig. 4.5.

*The advent of stiffeners in offshore structures is relatively recent, and few investigations exist on this subject. Stiffener effectiveness depends on the loading applied. For example, a single central annular stiffener has practically no effect on the SCF of a T joint subject to an in-plane bending load. However, the same stiffener reduces the SCF by a factor of nearly three in the same joint subjected to axial loading.*

*A limit exists to the effectiveness of stiffening, both concerning the cross-section of a stiffener and the number of stiffeners employed. This limit is naturally associated with the loading applied. Fig. 4.6 shows schematically the relations between the SSCF (stiffened joint stress concentration factor) and I (stiffener moment of inertia) and n (number of stiffeners).*

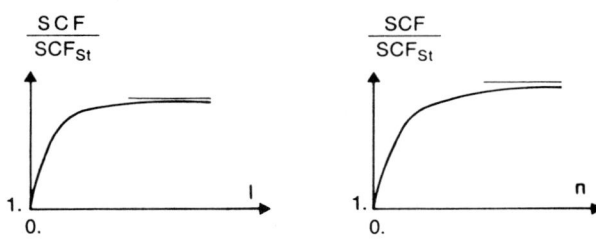

*Fig. 4.6.*

## 4.3.2 DETERMINATION OF THE SCF OF THE STIFFENED JOINT (SSCF)

At the present time, the SSCF is determined either experimentally or by a numerical method [ 4.9, 4.10, 4.11 and 4.12]. In other words, no parametric formulas are available to calculate the SSCF. A limited number of SSCF measurements exist on weld-fabricated joints stiffened by internal annular stiffeners. Fig. 4.7 shows the experimental measurements of the SSCF in comparison with SCF measurements found in the technical literature [4.13]. Apart from experimental measurements, calculations of the SSCF by finite elements exist for the same joints, confirming these measurements in most cases.

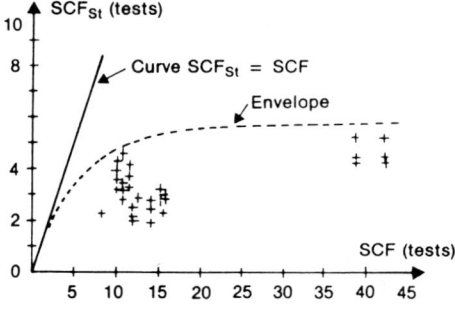

Fig. 4.7.

*No accurate methods exist for determining the SSCF from experimental measurements. The values given in Fig. 4.7 of the recommendations are those given by the authors of Refs. [4.9, 4.10, 4.11 and 4.12].*

## 4.3.3 FATIGUE STRENGTH OF STIFFENED JOINTS

Among the results of 18 fatigue tests conducted on stiffened joints, 12 were performed in Italy [4.12 and 4.14] as part of the ECSC programme, and 6 tests were performed in Japan [4.10]. Figure 4.8 gives the results of these 18 tests plotted on and S-N graph [4.13]. The abscissa shows the number of cycles $N_3$ corresponding to a through crack (see Section 5.1) for the Italian tests. As for the Japanese tests, the only result is $N_4$, the number of cycles corresponding to the end of the test.

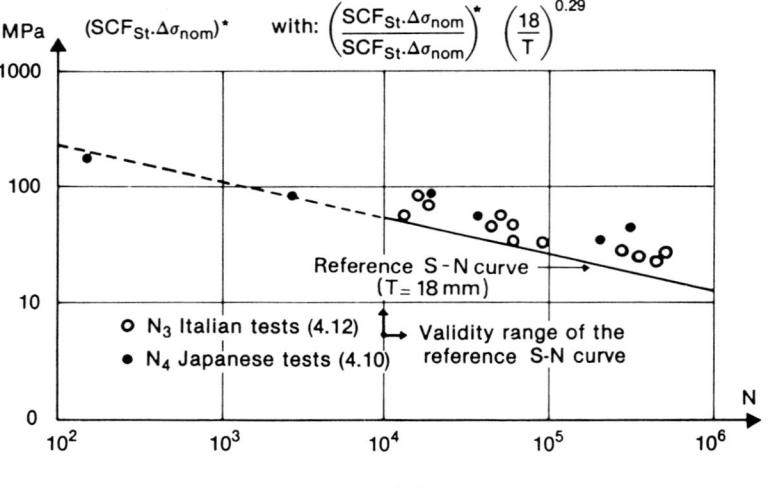

Fig. 4.8.

These few tests show the S-N curve for unstiffened tubular joins may be used also for stiffened joints, since the fatigue test results on stiffened joints (criterion $N_3$) lie above the reference curve, which is based on the $N_3$ criterion, for N $\geq 10^4$ cycles.

Note that the ordinate corresponds the product of the SCF multiplied by the nominal stress range.

*These 18 test results on stiffened joints cover a restricted validity range and do not provide grounds to assert positively that the S-N curve (see Chapter 5) is valid for large numbers of cycles. Furthermore, the fact should be noted that the position of the hot spot in stiffened joints is not the same as in unstiffened joints. In fact, the crack does not necessarily start at the toe of the brace-chord weld. It may initiate at the toes of the welds connecting the stiffeners.*

## 4.3.4  EFFECTS OF STIFFENING

Apart from the fact that it depends on the inertia and position of the stiffeners, the effect of stiffening is closely dependent on the type of loading applied. In actual fact, a stiffener simultaneously:

(a) Reduces the stresses in the joint due to a local addition of material.

(b) Locally increases the inertia of the chord due to the rigidity provided by the stiffener.

However, in the current state of the art, it is difficult to quantify each of these two effects. Some authors propose estimating the SSCF from parametric formulas by altering the parameter $\tau = t/T$ to account for the effect of the local addition of material and the parameter $\gamma = D/2T$ to account for the effect of local increase in inertia of the chord.

These approaches only offer an order of magnitude of the SSCF. For the time being, the values of the SSCF being low, an assessment of the accuracy of this approach is not possible.

*A method based on the use of parametric formulas exists for evaluating the stress concentration factor in simple stiffened joints [4.15].*

*For a stiffener acting on a length P, defined subsequently (see Fig. 4.10), the parameter $\tau_{eq}$ is defined by:*

$$\tau_{eq} = \frac{\tau}{K_1} \quad \text{with} \quad K_1 = \frac{S_{eq}}{S}$$

*where S and $S_{eq}$ are the longitudinal cross-section of the chord over the length P without and with stiffeners respectively (see Fig. 4.9).*

The second effect is taken into account by the parameter $\gamma = D/2T$. This parameter implicitly reflects the rigidity of the chord. $\gamma$ must therefore be modified to represent the local increase in inertia provided by the stiffeners. The coefficient $\gamma_{eq}$ is defined as follows:

$$\gamma_{eq} = \frac{\gamma}{K_2} \qquad \text{with} \quad K_2 = \left\{ \sqrt[3]{\frac{I_{eq}}{I}} \right\}$$

where $I$ and $I_{eq}$ are the inertias of the chord cross-sections without and with stiffeners respectively. $K_2$ is a cubic root, because for a rectangular cross-section, the thickness T appears at the 3rd power.

$S = P.T$

$I = P.T^3/12$

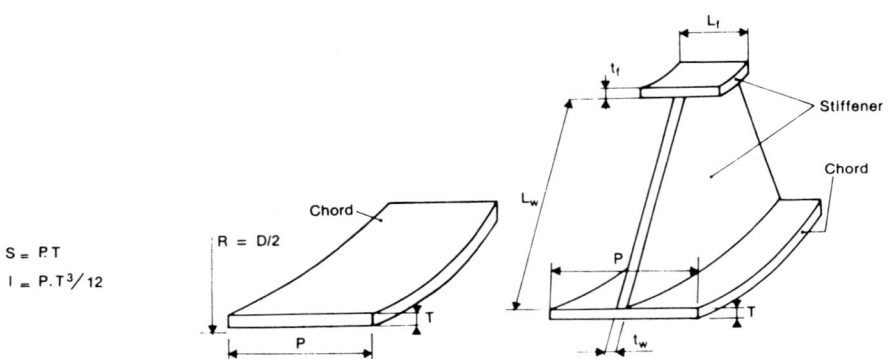

Fig. 4.9.

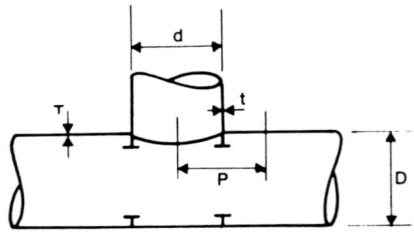

Fig. 4.10.

*Based on comparative studies, certain authors* [4.15] *propose taking* $K_2 = \left\{\sqrt[3]{\dfrac{I_{eq}}{I}}\right\}^{\eta}$ *with:*

$\eta$ = 1.00 *for the axial load,*
$\eta$ = 0.85 *for out-of-plane bending load,*
$\eta$ = 0.30 *for in-plane bending load.*

*It remains to define P, the length of the chord on which the stiffener acts. Following comparisons with the experimental results, the same authors* [4.15] *propose taking* $P = 1.56 \sqrt{RT}$, *where R is the radius and T its thickness. In the reference quoted, it is pointed out that for annular stiffeners, P = d gives slightly more conservative results (see Fig. 4.11).*

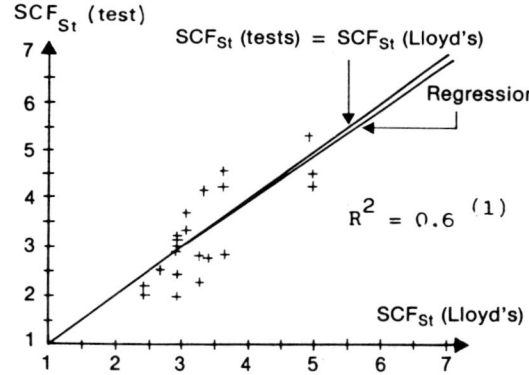

(1) $R^2$: *multiple correlation coefficient*

*Fig. 4.11.*

*The method for calculating the SSCF from modified parameters* $\tau_{eq}$ , $\gamma_{eq}$ *and from standard parametric formulas has only been confirmed for a number of specific cases. Although it is difficult for the time being to judge its validity, it can nevertheless be used for a preliminary project design.*

*Figure 4.11 gives the relation between the measured SSCF and the SSCF calculated by Lloyd's parametric formulas* [3.7] *by the approach discussed above with:*

(a) $P = 1.56 \sqrt{RT}$, *if P is less than the distance between two stiffeners.*

(b) *If not P = d.*

## REFERENCES

4.1 Mézière, Y., Etude d'un joint en K avec recouvrement, ANMT Report No.19, Laboratoire de Mécanique des Solides, September 1982.

4.2 Moe, E.T. and Gibstein, M.B., Stress analysis and fatigue failure of K-joints with overlapping braces, DnV Report No.80-1172, December 1980.

4.3 Gulati, K.C., Wang, W.J. and Kan, K.K.Y., An analytical study of stress concentration effects in multibrace joints under combined loading, OTC, Paper No.4407, Houston, Texas, 1982.

4.4 Moe, E.T. and Gibstein, M.B., Stress analysis and fatigue failure of K-joints, DnV Report No.80-1127, December 1980.

4.5 Gérald, J. and Mézière, Y., Etude d'un joint en K, ANMT Report No.16, Laboratoire de Mécanique des Solides, July 1980.

4.6 Recho, N. and Brozetti, J., Concentration de contraintes dans les piquages de tubes due à des sollicitations axiales, CTICM Report No.10.002.6 (confidential document with restricted circulation), May 1982.

4.7 Wordsworth, A.C. and Smedley, G.P., Stress concentrations at unstiffened tubular joints, European Offshore Steel Research Seminar, Cambridge, November 1978.

4.8 Dijkstra, O.D. and De Back, J., Fatigue strength of tubular X- and T-joints (Dutch tests), ECSC/IRSID International Conference, Steel in Marine Structures, Paper No.8.4, Paris, October 1981.

4.9 Salama, M.M., Optimization of stiffening rings, Position in welded tubular X-joints, Production Research Division, Conoco Inc., Ponca City, Oklahoma.

4.10 Sawada, Y. Idogaki, S. and Sekita, K., Static and fatigue tests on T-joints stiffened by an internal ring, Offshore Technology Conference, OTC, Paper No.3422, 1979.

4.11 Shiyekar, M.R., Kalani, M. and Belkune, R.M., Stresses in stiffened tubular joints of an offshore structure, Proceedings of the First Offshore Mechanics/Deep Sea System Symposium, New Orleans, 7/10 March 1982.

4.12 Pozzolini, P.F., Tests on tubular joints, ECSC/IRSID International Conference, Steel in Marine Structures, Paris, October 1981.

4.13 Recho, N. and Brozzetti, J., Présentation générale des problèmes relatifs aux noeuds raidis, ARSEM/CTIM report No.10.002.12, September 1983.

4.14 Brandi, R., Behavior of unstiffened and stiffened tubular joints, ECSC/IRSID International Conference, Steel in Marine Structures, Paris, October 1981.

4.15 Complex joint and loading, Fatigue study, Final report, Lloyds Register of Shipping, Offshore Services Group, February 1983.

# CHAPTER 5

# Definition
# of the Reference S-N Curve

## 5.1 DEFINITIONS AND BASES OF THE DETERMINATION OF THE S-N CURVE

The reference S-N curve relates the variation in design stress (see Chapter 1) expressed in N/mm² to the number of cycles N characterising the failure of a joint.

With respect to the behavior of a real platform joint, it may be considered that the through crack criterion presents a conventional critical failure state.

*By the use of specific inspection techniques, the through crack can be detected in a real structure. Thanks to structural redundances, the appearance of a through crack in any joint of a "jacket" in no way signifies the collapse of the structure, but it nevertheless consitutes a primary undermining of the integrity of the structure. The structure may no longer be able to perform the functions for which it has been designed with the same degree of safety.*

## 5.2  REFERENCE S-N CURVE AND VALIDITY CONDITIONS

The reference S-N curve is expressed as follows:

(a) For $N \leq N_C = 10^7$:

$$\log N = 12.29 - 3.00 \log S$$

(b) For $N > N_C = 10^7$:

$$\log N = 15.82 - 5.00 \log S$$

The plot of this reference S-N curve is given by the bi-logarithmic diagram in Fig. 5.1.

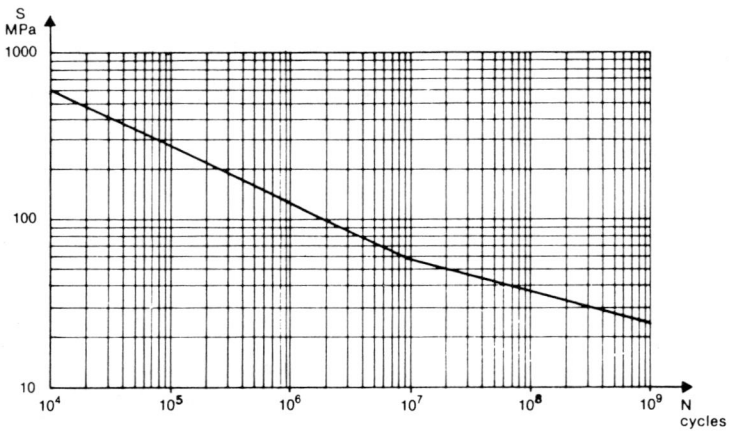

Fig. 5.1. Reference S-N curve for T = 18 mm

This curve corresponds to:

(a) A chord thickness T = 18 mm (use of the S-N curve for T ≠ 18 mm is indicated in Chapter 6).

(b) A survival probability of about 97.5%, the standard deviation of log N being 0.275.

(c) Loads applied to the structure in air, or protected against the effects of corrosion (see Chapter 4, Part I).

(d) As-welded joints.

*The reference S-N curve was established from the statistical analysis of fatigue tests on tubular joints of the European Programme of ECSC on the "fatigue behavior of offshore welded steel structures". A close examination of the test results led to the discarding of several tests so as to obtain a homogeneous sample. This meant that only 73 tests were considered out of more than 250. A statistical treatment of this sampling is discussed in detail in Ref. [5.1].*

*The total sample of 73 tests met the following conditions:*

*(a) The test was conducted under simple loading of constant amplitude, in the air.*

*(b) The geometric parameter $\beta$ (=d/D) was always below 1.0.*

*(c) The through crack characterised by the number of cycles $N_3$ occurred in the chord, at the weld toe.*

*(d) The measured values of S and $N_3$ were available.*

*The values of $N_3$ vary in the sample between $10^4$ cycles and $3 \times 10^7$ cycles.*

*The sample exhibits a significant scatter of the results, which can be attributed to the different chord thickness (T). The sample was divided into four groups for the four thickness ranges (T): 5-10, 15-25, 30-45 and 70-80 mm.*

*As for the 4th group (thickness 70-80 mm), the sample size was insufficient to plot a representative curve.*

*The statistical tests on the three other groups are given at Table 5.1.*

*All the tests of group 1 were performed on small joints, with chord diameters between 168 and 170 mm. In these conditions, the weld is not to the right scale. For these reasons, group 1, not being representative of the joint dimensions used in offshore structures, was discarded for establishment of the reference S-N curve. Therefore, the remaining two groups, 2 and 3, only were used for this purpose. The foregoing table shows that group 2 and group 3 display closely comparable slopes. This slope was imposed for group 4. The results of statistical analysis with the imposed slope of -3.00 are given in the Table 5.2.*

*Table 5.1.*

| Group | T (mm) | T (mm) mean | Size | Mean of regression[1] | | Standard deviation | Statistical tests | | |
|-------|--------|-------------|------|---------|---------|---------|---------|---------|---------|
| | | | | $A_1$ | $A_2$ | $A_3$ | $R^2$ | F-test | Correl. coef. |
| 1 | 5-10 | 6.3 | 22 | 12.13 | -2.50 | 0.43 | 0.51 | 40.9 | 0.72 |
| 2 | 15-25 | 18 | 18 | 12.84 | -3.00 | 0.25 | 0.895 | 157.4 | 0.95 |
| 3 | 30-45 | 34 | 27 | 12.57 | -3.01 | 0.29 | 0.80 | 125.3 | 0.89 |
| 4 | 70-80 | 76 | 6 | | N o t   s i g n i f i c a n t | | | | |

(1) *Estimation of the mean:* $Log N = A_1 + A_2 Log S$.
     $R^2$: *Multiple correlation coefficient.*

*Table 5.2.*

*Slope $A_2 = -3$ imposed*

| Group | T (mm) | T (mm) mean | Size | Mean of regression | | Standard deviation | Statistical tests | | |
|-------|--------|-------------|------|---------|---------|---------|---------|---------|---------|
| | | | | $A_1$ | $A_2$ | $A_3$ | $R^2$ | F-test | Correl. coef. |
| 1 | 5-10 | 6.3 | 22 | | | | | | |
| 2 | 15-25 | 18 | 18 | 12.84 | -3.00 | 0.25 | 0.895 | 157.4 | 0.95 |
| 3 | 30-45 | 34 | 27 | 12.55 | -3.00 | 0.30 | 0.79 | 125.3 | 0.89 |
| 4 | 70-80 | 76 | 6 | 12.34 | -3.00 | 0.55 | | | |

The reference S-N curve was finally based on the following considerations (decision of the ARSEM Technical Committee):

(a) The reference S-N curve corresponds to the mean curve of group 2 (thicknesses 15-25 mm, mean thickness 18 mm) less twice the generalised standard deviation.

(b) The generalised standard deviation is taken as the mean of the standard deviations of the two groups 2 and 3.

### Choice of change in slope

*Very few test results are available for which failure occurred after $10^7$ cycles.*

*Calculations based on the concepts of fracture mechanics (see Chapter 9) [5.2 to 5.4] show that in air, the S-N curves change slope at about $5.10^6$ cycles, according to the choice of the initial defect size and the value of $\Delta K_S$ corresponding to the propagation threshold.*

*For the time being, there is no formal justification for the values $N_C = 10^7$ and slopes $m = -5$ which were adopted. Note that the presence of a horizontal endurance limit leads to a greater sensitivity, as concerns the damage calculated, to the class subdivision of the stress histogram (see Chapter 8).*

### Other S-N curves

*Other S-N curves exist, including those in Refs. [5.5 to 5.11]. It is the designer's duty to make sure of the consistency as regards the safety of the calculations when one of these curves is used instead of the one recommended in this guide.*

## REFERENCES

5.1    Recho, N. and Ryan, I., Etablissement des courbes S-N de référence pour les joints tubulaires, CTICM Report No.10.002.7, October 1982.

5.2    Amiot, P. and Radenkovic, D., Prévision de la durée de vie des noeuds tubulaires sous chargement d'amplitude constante, ECSC/IRSID International Conference, Steel in Marine Structures, Paris, October 1981.

5.3    Gurney, T.R., Cumulative damage calculations taking account of low stresses in the spectrum, Welding Research International, 6 (2) 51-76, 1976.

5.4    Tilly, G.P. and Nunn, D.E., Variable amplitude fatigue in relation to highway bridges, Proceedings of the Institute of Mechanical Engineers, Vol.194, No.27, 1980.

5.5    Rules and Regulations for the Construction and Classification of Offshsore Plateforms, Bureau Veritas, 1975.

5.6    Structural Welding Code, American Welding Society, ANSI/AWS D1.1-81, 1981.

5.7    API Recommended Practice for Planning, Designing and Construction of Fixed Offshore Platforms, American Petroleum Institute, API.RP.2A, 12th Edition, 1981.

5.8    Code of Practice for Fixed Offshore Structures, British Standards Institution, BS 6235:1982.

5.9    Offshore Installations, Guidance on Design and Construction, Recommendations of Revision Drafting Panel, Issue G, Department of Energy, United Kingdom, March 1982.

5.10   Regulations for the Structural Design of Fixed Structures on the Norwegian Continental Shelf, Norwegian petroleum Directorage, 1977.

5.11   Rules for the Design, Construction and Inspection of Offshore Structures, Appendix C, Steel Structures, Det Norske Veritas, 1977.

# CHAPTER 6

# Modifications
# of the Reference S-N Curve

## 6.1 SCALE EFFECT

The scale effect has been attributed to the chord thickness T. The stress variation S concerning the reference S-N curve (see Chapter 5) is therefore modified as a function of T by the following equation:

$$S = S* \left(\frac{T}{18}\right)^{0.29} \qquad (T \text{ in } mm)$$

where S* is the value of the stress range calculated for the joint in question.

This is equivalent to the use of a modified S-N curve which is a function of thickness as follows:

(a) For $N \leq N_c = 10^7$ and $T \geq 15$ mm:

$$\log N = A(T) - 3.00 \log S*$$

with $A(T) = 13.38 - 0.87 \log T$.

(b) For $N > N_c = 10^7$ and $T \geq 15$ mm:

$$\log N = A(T) - 5.00 \log S*$$

with $A(T) = 17.64 - 1.45 \log T$.

For $T < 15$ mm, the curve corresponding to $T = 15$ mm is used.

Figure 6.1 gives the variation in A(T) as a function of T.

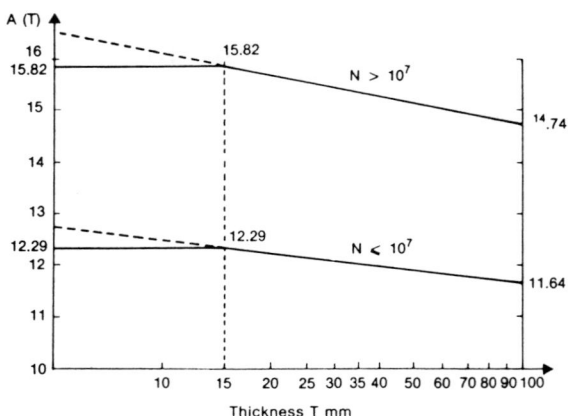

Fig. 6.1.

*The statistical analyses establishing the S-N curves for the three thickness groups are summarised in Table 6.1 [6.1].*

Table 6.1.

| Group | T (mm) | $T_j$ mean thickness | Mean line regression constant | | Standard deviation | Genera- lised standard deviation | Mean - 2 generalised standard deviation | | Size of data base |
|---|---|---|---|---|---|---|---|---|---|
| | | | $A_1$ | $A_2$ | $A_3$ | | $A_1$ | $A_2$ | |
| 2 | 15–25 | 18 | 12.84 | −3.00 | 0.25 | 0.275 | 12.29 | −3.00 | 18 |
| 3 | 30–45 | 34 | 12.55 | −3.00 | 0.30 | | 12.00 | −3.00 | 27 |
| 4 | 70–80 | 76 | 12.24 | −3.00 | 0.55 | | 11.79 | −3.00 | 6 |

*These three curves shown in Fig. 6.2, which correspond to the mean less twice the generalised standard deviation, have a slope of -3 for $N < 10^7$. The relation between S and T was obtained by linear regression on the three values of S, as a function of the mean thickness, for $N = 2.10^6$ cycles.*

*This thickness effect was extended to the values of N greater than 10[7].*

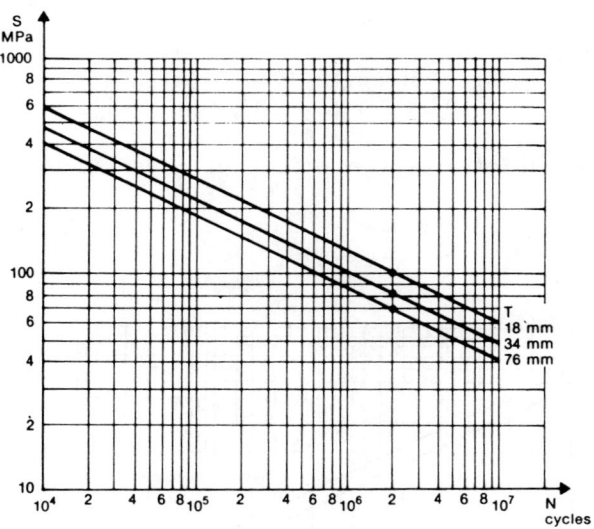

*Fig. 6.2.*

## 6.2  EFFECT OF CORROSION

The reference S-N curve (see Chapter 5) is used for joints protected against corrosion (see Chapter 4, Part I).   For joints subject to corrosion, the fatigue strength is estimated to be half of that of the joint in air for N $\leqq 10^7$.  For N $> 10^7$, the change in slope is ignored.

Figure 6.3 illustrates the effect of these changes on the reference curve.

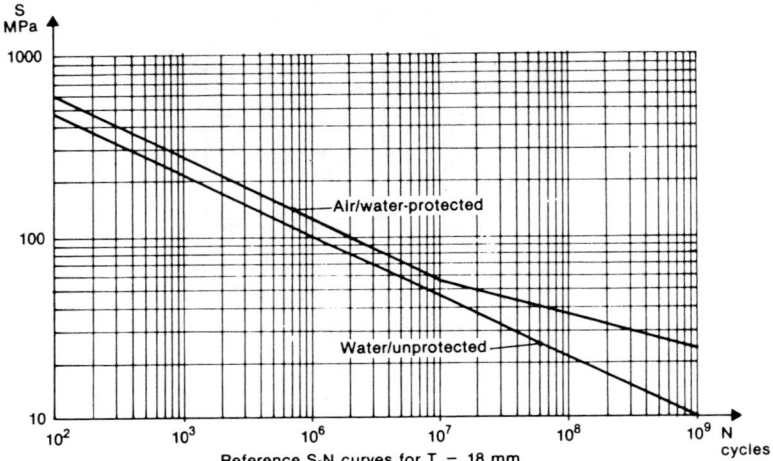

Fig. 6.3. Reference S-N curve for T = 18 mm.

*It is generally acknowledged that joints with cathodic protection have a service life in seawater close to that of joints in air [6.2].  However, joints unprotected against corrosion have a service life of about half similar joints in air for $10^4 < N < 10^7$.*

*As for joints subject to large members of cycles $N > 10^7$, the few tests conducted appear to indicate that the bend characterising the change of slope in the S-N curve does not exist in a corrosive environment.*

## REFERENCES

6.1   Recho, N. and Ryan, I., Etablissement des courbes S-N de référence pour les joints tubulaires, CTICM Report No.10.000.7, October 1982.

6.2   De Back, J., Strength of tubular joints, Plenary Session 7, ECSC/IRSID International Conference, Steel in Marine Structures, Paris, October 1981.

CHAPTER **7**

# Fatigue Strength
# Improving Techniques

### 7.1 INTRODUCTION

The fatigue strength analysis of the structure helps to identify a number of critical joints. To enhance the possibility of good fatigue behavior for these joints, the following improvement and finishing techniques need to be considered:

(a) Improved weld toe geometry by the use of a suitable welding procedure, in particular by the use of special electrodes (see Section 7.3).

(b) Remelting of the weld toe by means of TIG or plasma dressing (see Section 7.4).

(c) Weld toe grinding or local machining (see Section 7.5).

(d) The introduction of compressive residual stresses at the weld toe region by superficial prestressing: hammer and shot peening (see Section 7.6).

(e) Stress relieving heat treatment (see Section 7.7).

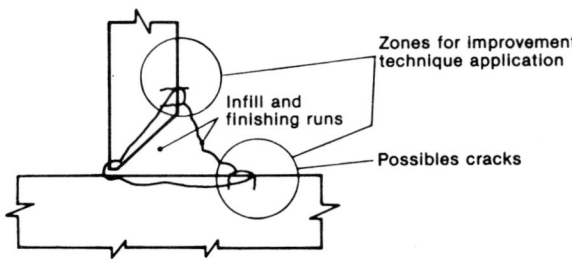

Fig. 7.1.

By means of the first three techniques, the conditions set by the acceptability criterion corresponding to the execution of the controlled profile weld can be met (see Section 7.2). Hammer and shotpeening only slightly alter the shape of the profile.

On tubular joints, the first welding runs can be preferential zones of fatigue crack initiation. Treatment by an improvement technique must above all attempt to cover the weld toe region (see Fig. 7.1).

*The principles underlying improvement techniques are as follows:*

*(a) The increase or the introduction of the crack initiation phase, by altering the notch geometry or by eliminating defects.*

*(b) Alteration of the effective local stress field, by introducing beneficial superficial residual compressive stresses, or by stress relieving heat treatment of the welded joint leading to the relaxation of the residual tensile stresses induced by the welding process.*

*The test results show that in a corrosive environment, improvement techniques achieve lesser gains in fatigue life than in the open air, but they are nevertheless significant. However, if an improvement technique is combined with cathodic protection to eliminate pitting corrosion, the initiation period is likely to be increased in these conditions and should correspond to a significant part of the total fatigue life. Other experimental investigations are necessary to confirm these observations, before drawing any final conclusions.*

*Few results are at present available concerning the effectiveness of improvement techniques on tubular joints. However, significant results were obtained on small test specimens, which need to be confirmed on larger specimens [7.1 to 7.5].*

*The user's attention is drawn to the difficulty of quantifying the gain in the fatigue strength of a joint to which a specific improvement treatement has been applied. In other words, especially for the time being, the results of these different techniques are qualitative rather than quantitative.*

## 7.2  ACCEPTABILITY CRITERION FOR A CONTROLLED PROFILE WELD

The rules concerning the weld profile in the neighborhood of the weld root are specified in Chapter 2, Part I.  To set the conditions for inspection of the very local execution of the weld toe transition, the acceptability criterion defined in Fig. 7.2 can be employed.

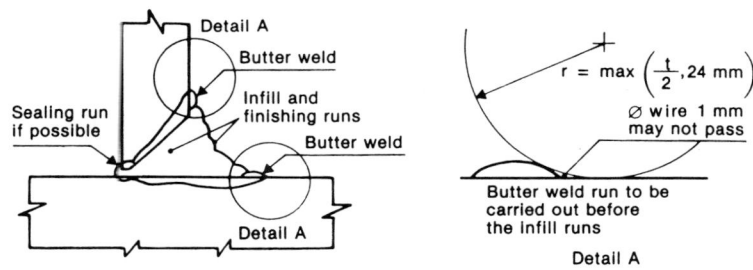

$$r = \max\left(\frac{t}{2}, 24 \text{ mm}\right)$$

Ø wire 1 mm may not pass

Butter weld run to be carried out before the infill runs

Detail A

Fig. 7.2.

*The weld toe is a sensitive zone in which defects such as undercuts may occur, as well as a notch effect.  These defects reduce or even nullify the part of service life corresponding to the crack initiation period (Fig. 7.3).*

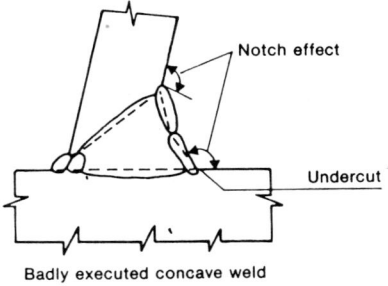

Badly executed concave weld

*Fig. 7.3.*

## 7.3 IMPROVEMENT OF THE WELD TOE BY THE USE OF SUITABLE WELDING PROCEDURES

The use of special electrodes helps to guarantee a progressive transition of the connection geometry at the weld toe, and thus reduces the local stress concentration. Two techniques are commonly employed:

(1) The use of a special electrode with improved flow characteristics.

(2) Execution of an additional run with a special electrode with improved flow characteristics.

*Both techniques have been used mainly in Japan [ 7.6 to 7.9 ]. The first technique (use of improved flow electrodes) demandes no additional work after welding. It is the choice of the electrode, with a suitable coating, that serves to obtain a very progressive geomtric transition between the weld and the tube wall, at least for ,a flat position weld. This is due to the good flow characteristics of the filler metal.*

*The second technique consists of the execution of one additional run with an improved flow electrode. This must not be confused with a buttering weld run (see Chapter 2, Part I).*

*It is improbable that the good results obtained for a horizontal or flat position weld are valid for all welding positions.*

## 7.4   WELD TOE REMELTING

### 7.4.1  TIG DRESSING

The technique consists of the remelting of the weld transition zones by means of a TIG torch. This results in a decrease in the local concentration by flattening the weld profile at the transition, and elimination by fusion of the weld toe defects [ 7.10 to 7.16 ].

Surface cleaning helps to eliminate scale, slag and other impurities liable to cause small notches or undercuts directly at the new transitions. It is recommended to perform preliminary tests to determine the optimal protection gas flow rate. This helps to provide a stable arc (an unstable arc may cause defects, such as holes in the weld, or oxidation of the electrodes [ 7.17 ]).

*The best results are obtained when the TIG improvement pass in executed without a filler metal and when the heat input is at least to 10 kJ/cm.*

*A properly spread stable arc, covering a wide area, is necessary for the metal to be remelted in a wide enough zone and to obtain satisfactory profiles. The improvement depends on the position of the torch and the best results are obtained when the arc is directed towards the base metal at a distance of 0.5 to 1.5 mm from the weld toe.*

*The improvement depends on the joint geometry. Moreover, it appears that the improvement increases with the yield strength of the base metal.*

*In some steels, TIG dressing causes an increase in hardness in the heat affected zone (HAZ). Hardnesses of around 400 HV are often obtained. Certain regulations do not allow hardnesses over 250 HV in offshore structures (see Chapter 3, Part I). A second run, 3 to 4 mm from the first, causes tempering of the martensitic zone formed by the first run, and may restore the hardness of the weld toe down to an acceptable level.*

### 7.4.2 PLASMA DRESSING

Plasma dressing offers the following advantages over the TIG process:

(a) The hardness of the heat affected zone (HAZ) is lower, due to the higher energy input.

(b) The remelted zone at the weld toe is wider, giving a better profile.   The result of the operation is less sensitive to the position of the torch in relation to the weld root.

(c) The travel speed is higher due to the greater energy input.

(d) Electrode maintenance is minimal.

(e) According to published results, improvements in fatigue strength are slightly better.

Use of this technique also incurs the risk of an unfavorable increase in the hardness of the steel in the weld zone, but not as much as would TIG dressing.


   *Plasma dressing employs a torch and electrode that are different from those used for TIG dressing [7.18, 7.19].*

   *The energy input rate is about 50 to 100% greater for plasma dressing than for TIG dressing.*

## 7.5  WELD TOE GRINDING OR MACHINING

The technique consists of grinding the weld toe to a depth of 0.5 to 0.8 mm, either with a grinding disc or a rotary burr grinder to eliminate incipient cracks, slag inclusions and micro cracks. The technique serves to improve the geometry at the weld toes and hence lower the local stress concentration factor [7.20 to 7.23].

The care taken in executing the operation largely conditions the results obtained.

*With disc grinding, the grinding disc grains leave machining scratches parallel to the weld. These scratches are potential initiation sites and it is preferable to eliminate them with a small burr grinder.*

*If a rotary burr grinder is used, the ground surface is often marked by chips adhering to the cutter, and this may have an unfavorable effect in terms of crack initiation.*

## 7.6  SUPERFICIAL PRESTRESS

The introduction of high compressive residual stresses in a superficial layer of the metal around the weld toe is achieved either by hammer peening or by shot peening at the weld toe.

> *The principle of this prestressing may raise some problems in its application to offshore structures. For many structures, in fact, the loading is such that the yield stress is reached locally (stress concentration). Offshore structures are subject to loading histories that are of variable amplitude and are also random. It is possible that high amplitude cycles may progressively alter the residual stress field introduced by hammer or shot peening. If so, the beneficial effect of such mechanical surface treatments is likely to be jeopardized, even if the majority of the stress cycles are of low amplitude.*

### 7.6.1  HAMMER PEENING

A pneumatic hammer is used to cold hammer the weld toe.  The hammer is fitted with a hard steel hemispherical head tool or a multipoint round head needle device [7.22 to 7.25, 7.30].

The effectiveness of hammer penning depends on the number of passes and the duration of the operation.  An identation depth of 0.6 mm, obtained in four passes of the hammer, is generally aimed at, which offers a good compromise between treatment time and effectiveness.

Outside of laboratory conditions, the difficulties of implementing this technique make the results rather haphazard.

> *After hammer peening, the defects are embedded in a layer of work hardened material in which high residual compressive stresses exist, induced by the hammering operation. Excessively rapid hammering is liable to give rise to a work hardened zone insufficiently deep to enclose all the defects in the residual compressive stress field.*

*Tests have shown that a small part of the improvement in fatigue strength obtained by hammer peening derives from a change in the weld toe geometry (the transition is better) and a change in the form of the existing defects.*

*Hammer peening with the hard steel hemispherical head tool causes deeper identation than with the needle device. This may explain a greater increase in fatigue life if the first procedure is employed. Hammer peening often achieves a greater improvement in fatigue strength than TIG, plasma and shot peening techniques.*

*Many authors have pointed out the random aspect of the process, especially when its implementation requires a substantial manual contribution.*

## 7.6.2  SHOT PEENING

Shot peening involves bombarding the surface with roughly spherical shot, the impact of the shot being similar to small hammer blows.

The aim is to introduce compressive residual stresses into a superficial layer in such a manner as that the stress levels, the stress distribution pattern and the depth of the stressed layer are reproducible. In order to achieve this, it is necessary to control all of the parameters of the process: the diameter and hardness of the shot, the bombardment time and energy, the travel speed of the gun, the angle of the gun to the surface being treated, the distance of the nozzle to the surface and the nozzle diameter [7.26 to 7.29].

*Compressive residual stresses greater than $0.5 R_e$ can be introduced by shot peening. The considerable differences between the amounts of improvement that have been observed for various types of test specimen can probably be attributed to the differences between the initial residual stresses introduced during welding. Given the high level of automisation generally associated with applications of this technique, one can hope for a greater regularity in the treatment than with techniques, calling upon considerable manual intervention.*

## 7.7  STRESS RELIEVING HEAT TREATMENT

The conditions under which stress relieving heat treatment proves necessary are specified in Section 3.7.4, Part I. The stress relieving heat treatment program must be covered by a qualification procedure to make sure of a sufficient relaxation of residual stresses and the guarantees related to steel properties. It is recommended to set up the stress relieving heat treatment program and the qualification procedure jointly with the steelmaker and the inspection organisation.

The following factors must be taken into account in the preparation and implementation of the process [ 7.30 to 7.31 ]:

(a) It is prefereable to treat the joint entirely in a stress relieving furnace, but if this is unfeasible, the weld connection zones can be treated individually.

(b) The layout of heating elements and insulation must be such that the temperature distribution profile must be more or less symmetrical about the central axis of the weld and uniform along its circumference.

*The beneficial effect of heat treatment relies on the relaxation of the tensile stress introduced by the welding process in the usual crack initiation zone, in other words, the weld toe.*

*If the joint is subject to entirely tensile stress cycles, the heat treatment is of little interest.*

*It is when at least part of the stress cycles correspond to compressive stresses that a significant improvement can be anticipated from a suitable treatment. For a joint of an offshore structure, the loading conditions are so variable and complex that it is rarely possible to determine the exact loading cycle throughout fatigue life (for each sea state, for example).*

*Published results reveal a substantial scatter in the degree of improvement that can be expected. This probably stems from differences between the stress relieving heat treatment processes and between the steel grades [ 7.32 to 7.34 ]. For example, the following stress relieving heat treatement was applied by IRSID to the French joints (E36-4, Z-35 steel) of the ECSC programme:*

*Post-weld heat treatment of the joint at 580°C; this temperature was reached at a temperature build-up rate of 50°C/h, and then held for a period of 100 to 200 min, depending on thickness.*

*Due to the uncertainties mentioned above (uncertainty concerning loading cycles, variety of procedures), the beneficial effect of heat treatment can only be determined qualitatively.*

## REFERENCES

7.1    Bignonnet, A., L'influence des traitements d'amélioration du pied de soudure sur la tenue à la fatigue des joints soudés, IRSID Report FA 3259, October 1981, IIW Doc.XIII 1085-83.

7.2    Haagensen, P.J., Improvement of the Fatigue Strength of Welded Joints, Plenary Session 6, ECSC/IRSID, International Conference, Steel in Marine Structures, Paris, October 1981.

7.3    Gurney, T.R., Fatigue of Welded Structures, Second Edition, Cambridge University Press, 1979.

7.4    Iida, K. and Ishiguro, T., Brief summary of Japanese documents concerned with improvement of fatigue strength of welded joints, IIW Doc.XIII 862-77.

7.5    Oliver, R. and Ritter, B., Improvement of Fatigue Strength of Welded Joins by different treatments    Statistical Analysis of Literature Data, ESCS/IRSID International Conference, Steel in Marine Structures, Paris, October 1981.

7.6    Kobayashi, K., Matsumoto, S., Tanaka, M., Funakoshi, T., Sakamoto, N. and Shinkawa, K., Improvement in the fatigue of fillet welded joint by use of the new welding electrode, IIW Doc.XIII 828-77.

7.7    Kanazawa, S., Ishigura, T., Hanzawa, M. and Yokota, H., The improvement of fatigue strength in welded high tensile strength steels, IIW Doc.XIII 735-74.

7.8    Todoroki, R., Hanzawa, M., Ishiguro, T. and Yanagimoto, S., Effect of toe profile improvement on corrosion fatigue properties of welded joints, IIW Doc.XIII 875-78.

7.9    Todoroki, R., Sekiguchi, S., Ishiguro, T. and Zaizen, T., Problems on improvement of corrosion fatigue strength of steel in sea-water, Metallic corrosion, 8th International Conference, Mainz, 1981.

7.10    Millington, D., TIG dressing for the improvement of fatigue properties in welded high strength steels, IIW Doc.XIII 698-73.

7.11    Kado, S., Ishiguro, T., Hanzawa, M. and Yokota, H., Influence of the conditions in TIG dressing on fatigue strength in welded high strength steels, IIW Doc.XIII 771-75.

7.12 Simon, P. and Bragard, A., Amélioration des propriétés de fatigue des joints soudés, CEC Agreement No.6210-45/2/202, Final report.

7.13 Minner, H.H. and Seeger, T., Investigation of the fatigue strength of weldable high strength steel St E460 and St E490 in as-welded and TIG-dressed conditions, IIW Doc.XIII 912-79.

7.14 Hanzawa, M., Yokota, H., Ishiguro, H., Takashima, H., Kado, S., Tanigaki, T. and Hashida, Y., Improvement of fatigue strength in welded high tensile strength steel by toe treatment, IIW Doc.XIII 829-77.

7.15 Haagensen, P.J., TIG dressing of steel weldments for improved fatigue performance, OTC, Paper No.3466, April 1979.

7.16 Booth, G.S., Constant amplitude fatigue tests on welded steel joints performed in air, European Offshore Steel Research Seminar, Cambridge, November 1978.

7.17 The method of TIG dressing (Anon.), Welding in the World, Vol.14, Nos.3/4, 1976.

7.18 Kado, S.W., Ishiguro, T. and Ishii, N., Fatigue Strength improvement of welded joints by plasma arc dressing, IIW Doc.XIII 774-75.

7.19 Shimada, W., Hoshinouchi, S., Hiramoto, S., Hisikata, A., Yoshioka, S. and Inoje, A., Improvement of fatigue strength in fillet welded joint by $CO_2$ soft plasma arc dressing on weld toe, IIW Doc.XIII 881-78.

7.20 Schofield, K.G., Improving the fatigue strength of fillet welded joints by disc grinding the weld toe, Welding Institute Member's Report E/60/75, 1975.

7.21 Mullen, C.L. and Merwin, J.E., Fatigue life improvement by weld reinforcement and toe grinding, OTC, Paper No.4240, Houston, Texas, 1982.

7.22 Knight, J.W., Improving the fatigue strength of fillet welded joints by grinding and peening, Welding Institue Member's Report 8/1976/E.

7.23 Booth, G.S., The fatigue life of ground or peened fillet welded steel joints, Metal Construction, 13, 1981.

7.24  Faulkner, M.G. and Bellow, D.G., Improving the fatigue strength of butt welded joints by peening, Welding Research International, 5 (3), 1975.

7.25  Recommandations pour l'Application d'un Traitement au Marteau à Aiguilles, IS Document 167-65.

7.26  Shot-peening of Metal Parts, US Military Specification MIL-S-13165B, Amendment Z, 25 June 1979, reprinted by Metal Improvement Company Inc.

7.27  Shot-peening Applications, Metal Improvement Company Inc, 6th Edition, 1980.

7.28  Flavenot, J.F. and Niku-Lari, A., Le grenaillage de précontrainte, Etude bibliographique, CETIM Technical Note No.15, 1976.

7.29  Minutes of the First International Conference on Shot-peening organized by CETIM, Paris, September 1981.

7.30  Mécanosoudage-Fabrications, CETIM, 1983.

7.31  Code of Practice for Fixed Offshore Structures, British Standards Institute, BS 6235:1982, Section 6.8.8.

7.32  Sanz, G., Lieurade, H. and Gérald, J., Fatigue tests on Ten Full Scale Tubular Joints, ECSC/IRSID Conference, Steel in Marine Structures, Paris, October 1981.

7.33  Lourenssen, A., and Dijkstra, O.D., Fatigue tests on large post weld heat treated and as-welded tubular joints, OTC, Paper No.4405, Houston, Texas, 1982.

7.34  Shinners, C.D. and Abel, A., Fatigue of as-welded and stress relieved tubular T-joints, AIPC Symposium, Lausanne, 1982.

CHAPTER **8**

# Cumulative Fatigue Damage

This Chapter is devoted to the problem of fatigue, in other words, damage incurred in a welded joints by successive stress cycles.

To assess the fatigue strength of a joint, the designer has to determine the joint fatigue life, based on stresses applied to the joint in time (see Chapter 2), increased by the stress concentration factors (see Chapters 3 and 4) and on the relevant fatigue strength or S-N curve (see Chapters 5 and 6). This calculation is treated in this Chapter.

## 8.1  METHODOLOGY

The techniques discussed here serve to determine the fatigue life of a joint subject to known loads in time.  In addition, the calculation allows one to determine a value D called the damage factor, which depends on time, and is by definition 1 for the calculated fatigue life.

1. Fatigue life is an observable quantity.  It is assumed that after the calculated fatigue life has elapsed, the joint fails by the through cracking criterion defined in Chapter 5.

The damage factor D is by definition 0 if the joint has not yet been subjected to loads, and 1 if the joint has failed.  Between these two limits, it is defined by means of a cumulative damage rule written as follows:

$$D(T) = \Sigma \; \frac{n(S)}{N(S)}$$

where

n(S) is the number of cycles at a stress range of S applied to the joint up to the time T considered, and

N(S) is the number of cycles at a stress range of S causing failure of the joint, which is intact before the application.

*Cumulative damage rules can account for different parameters such as sequence effects, material properties, etc.  For offshore structures, the state of knowledge and the relative absence of experiments have led to the use of a linear relation.  This rule does not account for the chronology of the cycles or of the value of the mean stress at each cycle.*

2. N(S) is given by the S-N curve defined in Chapters 5 and 6.

n(S) is determined from the results of load calculations and of the stress concentration calculations, described in Chapters 2, 3 and 4.

S represents a geometric stress range in time.  The geometric stress $\sigma_G$ in time may be available in three different forms:

(a) A histogram of stress ranges S of the geometric stress $\sigma_G$ (Fig. 8.1).

(b) A set of spectral density functions of the geometric stress process (Fig. 8.2), for a number of short-term load states, each assigned a probability of occurence.

(c) A set of trajectories of the geometric stress process (Fig. 8.3) for a number of short-term load states, each assigned a probability of ocurrence.

Depending on each specific case, n(S) is then read, calculated or counted.

The damage factor that can then be calculated corresponds to the calculation time considered.  The fatigue life is determined by dividing the calculation time by the calculated damage factor.

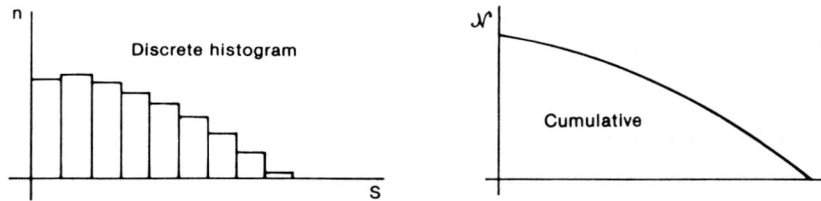

Fig. 8.1. Histogram of geometric stress ranges.

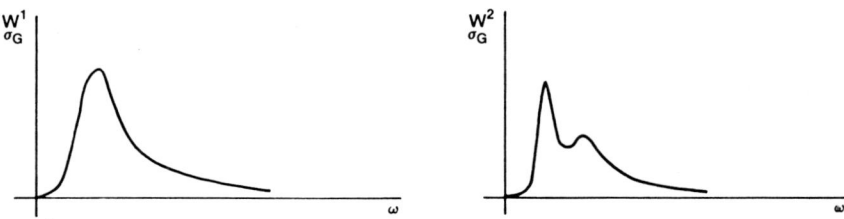

Fig. 8.2. Spectral density function of short-term loading states. (Each spectrum corresponds to a given short-term sea state).

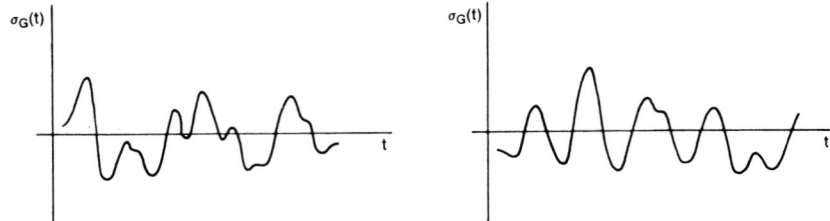

Fig. 8.3. Trajectories of a geometric stress process.
(Each trajectory corresponds to a given short-term
sea state).

3. It should be noted that the calculated damage factor contains an uncertainty. This uncertainty arises from all the steps in the computation sequence:

(a) Modelling of environmental forces.

(b) Calculation of nominal loads from the forces.

(c) Calculation of the stress concentration in the joint.

(d) S-N curve (especially in the region with a large number of cycles).

(e) The cumulative damage rule.

*For example, a relative error of 10% in the value of a stress concentration factor causes an error of 30 to 50% in the estimation of joint fatigue life.*

*However, it is recalled that the reference S-N curve (see Chapters 5 and 6) corresponds to the mean less two standard deviations of the retained test results.*

## 8.2 CUMULATIVE DAMAGE RULE

### 8.2.1 DISCRETE FORMULATION OF THE PALMGREN-MINER RULE

This formulation is adapted to fatigue tests in which a specimen is subjected to a loading consisting of a number of blocks of cycles of constant stress range, the ith block being defined by the number $n_i$ of cycles of stress range $S_i$ (Fig. 8.4).

The damage factor is then written:

$$D = \sum_i \frac{n_i}{N_i}$$

where $N_i$ is determined from $S_i$ by means of the S-N curve.

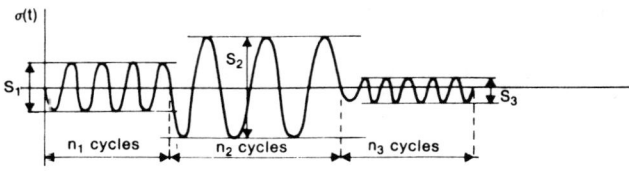

Fig. 8.4.

### 8.2.2 CONTINUOUS FORMULATION OF THE PALMGREN-MINER RULE

If the set of stress cycles S does not assume a discrete set of values, the damage law is written:

$$D(T) = \int_0^\infty \frac{n(S)dS}{N(S)}$$

where $n(S)dS$ is the number of cycles of stress ranges between S and S + dS during the calculation time T.

*In a constant amplitude cyclic loading test, the application of the Plamgren-Miner rule, which is reduced to $D = \frac{n}{N}$ assumes that the damage factor increases linearly with the number of cycles applied [8.1, 8.2, 8.3].*

*In a fatigue test by blocks, the application of this rule also assumes that the damage factor is obtained by adding the effects of the different blocks, without taking account of any interaction. This implies in particular that the order of application of the blocks is of no importance for the fatigue behavior of the member. This gives a continuous formulation of the rule for any loading case (each block consisting of a single cycle).*

*The following criticisms can be expressed concerning the Palmgren-Miner rule applied to any loading [8.4, 8.5]:*

*(a) The order of cycles is not taken into account.*

*(b) The existence of an endurance limit does not appear explicitly.*

*(c) The rule does not account for the scatter of results of tests at constant amplitude.*

*However, its simple formulation imposes the use of this rule for fatigue calculations on offshore structures. Note that the load calculation cannot provide any indication of the order of succession of stress cycles (Fig. 8.5).*

*Figure 8.6 describes other types of cumulative (nonlinear) rules qualitatively, for the case in which two blocks of stress cycles of scale $S_1$ and $S_2$ are applied in succession [8.6, 8.7].*

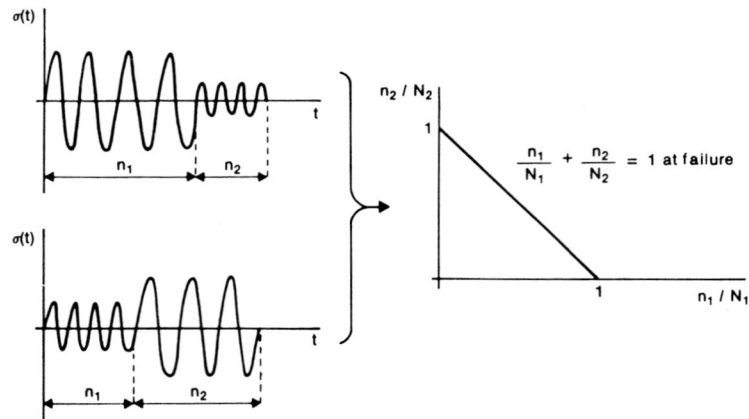

*Fig. 8.5. Palmgren-Miner rule.*

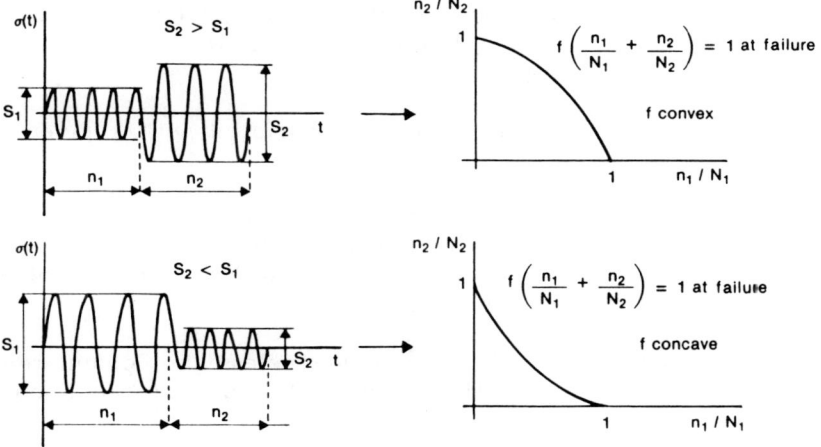

*Fig. 8.6. Nonlinear rule.*

## 8.3  RELATIONSHIP BETWEEN FATIGUE LIFE
## AND DAMAGE FACTOR

The damage factor calculated is associated with a given calculation time.

It should be noted that long-term statistics concerning forces (height-period relation of individual waves, significant height-mean apparent period relation of short-term sea states, etc.) must be established over a sufficiently long period (see Section 2.1).

The damage factor D can nevertheless be calculated for any calculation T. For any other time t, because of the linearity of the cumulative damage rule adopted, we have:

$$D(t) = \frac{t}{T} D(T)$$

The joint fatigue life, i.e. the time t for which D(t) = 1, is hence $\frac{T}{D(t)}$ irrespective of time T.

## 8.4 STRESS PROCESSING

This Section discusses how to determine n(S) from the stress calculation.

### 8.4.1 CASE WHERE A HISTOGRAM FOR S IS AVAILABLE

The interval of geometric stress ranges likely to be applied to the joint with a non-negligeable probability is broken down into p classes. For each class i, corresponding to $S_i \leqq S < S_{i+1}$, i = 1 to p, the histogram gives $n_i$ (number of cycles). One then calculates:

$$D = \sum_{i=1}^{p} \frac{n_i}{N(S_i^*)}$$

where $N(S_i^*)$ is the number of cycles to failure provided by the S-N curve.

It is recommended to take $S_i^* = 1/2 (S_i + S_{i+1})$, since the choice of $S_i^* = S_{i+1}$ is liable to lead to an excessively conservative estimate of D if the number of classes is not very large.

Similarly, instead of a stress range histogram, a cumulative of S and $N$ may be available, providing for each $S_i$, i = 1 to p, the number $N_i$ of cycles of stress range equal to or greater than $S_i$. This gives:

$$D = \sum_{i=1}^{p} \frac{N_i - N_{i+1}}{N(S^*)} \qquad \text{with} \quad N_{p+1} = 0$$

It is recommended that p should be sufficiently large. In particular, it is important to pay special attention to the fineness of the subdivision in the stress range region making greatest contributions to the damage factor.

*If the S-N curve has the equation $S^m N = A$, for* $S_i \leqq S < S_{i+1}$:

*then:*

$$S_i^* = \left( \frac{S_{i+1}^{m+1} - S_i^{m+1}}{(m+1)(S_{i+1} - S_i)} \right)^{1/m}$$

## 8.4.2 CASE WHERE A SET OF SPECTRAL DENSITY FUNCTIONS IS AVAILABLE

It is assumed that we have M spectral density functions of the geometric stress, each spectral density function $W_i(\omega)$ being associated with an occurrence time $t_i$ during the calculation time T. Similarly, it can be assumed that all the short-term stress states have the same duration t and that the number of occurrences $f_i$ during time T is known:

$$\sum_{i=1}^{M} f_i \, t = \sum_{i=1}^{M} t_i = T$$

Based on each spectral density function $W_i(\omega)$ one can define the spectral moments:

$$m_{i_k} = \int_0^\infty \omega^k \, W_i(\omega) \, d\omega \qquad k = 0, 2, 4, \ldots$$

and the parameter $\varepsilon_i$ of spectral width i:

$$\varepsilon_i = \sqrt{1 - \frac{m_{i_2}^2}{m_{i_0} \, m_{i_4}}}$$

The only case of interest here is the one in which the parameter $\varepsilon_i$ is sufficiently small ($\varepsilon_i < 0.4$) to be able to consider that the spectrum i is a narrow band spectrum. If not, refer to Section 8.4.3.

It is also assumed in this section, that the process whose spectral density function is known possesses all the properties mentioned in Section 2.1.4B, for the free surface elevation process and, in particular, that it is Gaussian.

The damage factor is expressed as a function of calculation time T by:

$$D(T) = \sum_{i=1}^{M} \frac{t_i}{2\pi} \sqrt{\frac{m_{i_2}}{m_{i_0}} \frac{1}{4 m_{i_0}}} \int_0^\infty \frac{S \, e^{-S^2/8 m_{i_0}}}{N(S)} \, dS$$

where N(S) is given by the S-N curve (see Chapters 5 and 6). The integral appearing in this expression of D(T) is generally evaluated numerically:

$$\int_0^\infty \frac{S e^{-S^2/8m_{i_o}}}{N(S)} \, dS = \sum_{i=1}^{p} \frac{S_k^* \, e^{-S_k^{*2}/8m_{i_o}}}{N(S_k^*)} (S_k' - S_k)$$

The stress interval of S is divided into p classes $[S_k, S'_k]$ (such that, for example, the number of cycles is the same for all classes). $S^*_k$ is selected equal to $1/2 (S_k + S'_k)$.

If the equation of the S-N curve has the form $S^m N = A$ where m is a whole number, the analytical expression of $D(T)$ can be determined.

*If the equation of the S-N curve has the form $S^m N = A$ (without slope change), the analytical expression of the damage factor is:*

$$D(T) = \frac{1}{A} 2^{\frac{3m}{2}} \Gamma\left(\frac{m}{2} + 1\right) \sum_i \left( \frac{t_i}{2\pi} \sqrt{\frac{m_{i_2}}{m_{i_o}}} \, m_{i_o}^{\frac{m}{2}} \right)$$

*with*

$$\Gamma(x) = \int_0^\infty u^{x-1} e^{-u} \, du \qquad \text{(the gamma function)}$$

## 8.4.3 CASE WHERE A SET OF TRAJECTORIES IS AVAILABLE

It is assumed here that M geometric stress trajectories are available, each trajectory corresponding to a steady state process. This geometric stress process may be Gaussian (wide-band) or non-Gaussian (if the behavior of the structure is nonlinear).

Each short-term state is characterised by a time of occurrence $t_j$ during the calculation time T, or similarly by a time t and a number of occurrences $f_j$ during time T.

The duration of each trajectory, which is independent in principle of t or $t_j$, must be sufficiently long to guarantee steady state conditions. It is also necessary to simulate each short-term stress state by means of several trajectories that are not inter-correlated.

For each available trajectory (j = 1 to M), the use of a counting method helps to determine, for each segment $S_i$, $S'_i$, the mean number of stress cycles per unit time with a stress range between $S_i$ and $S'_i$, or $n_{ij}$.

This helps to construct the S histogram. The number of stress cycles $n_i$ with stress range S between $S_i$ and $S'_i$ over the calculation time T is:

$$n_i = \sum_{j=1}^{M} f_j \, t \, n_{ij} = \sum_{j=1}^{M} t_j \, n_{ij}$$

Refer to Section 8.4.1 for the calculation of the damage factor D(T) from this histogram.

## Counting methods

The choice of a counting method depends on the way in which the stress cycles are defined. Starting with a definition of the cycle that is specific to it, each method proceeds to estimate the number of these cycles and their range S, for a time t of the trajectory analysed.

*The very concept of a cycle raises the problem of interpretation if the process is not sinusoidal and of constant amplitude in time. This explains the proliferation of counting methods.*

*References [ 8.10 to 8.12 ] offer a more detailed comparative study of the different methods.*

### "Rain-flow" counting method (a)

This method is used to determine stress cycles and half-cycles. Figure 8.7 provides one example. A detailed description of the method can be found in Ref. [8.10].

To use this method, it is therefore assumed that the number $n_i$ appearing in the expression of D in Section 8.4.1 may be fractional (i.e. half-cycles are counted).

### Peak counting method (b)

This method is used to count the positive peaks and negative lows, the other extremes being ignored. Each positive stress peak is associated with a cycle of stress range S equal to twice the peak amplitude (Fig. 8.8).

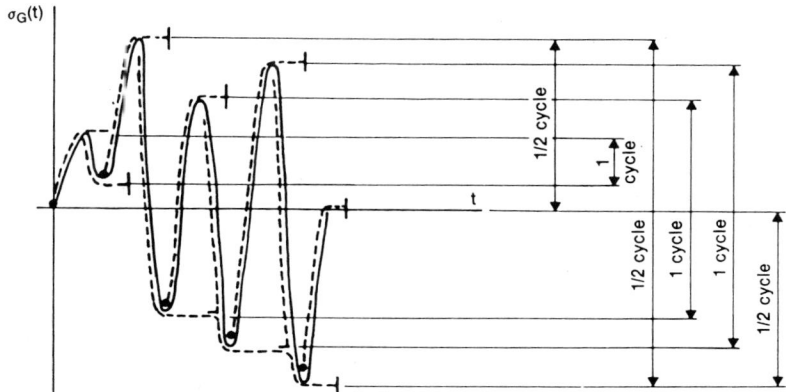

Fig. 8.7. Rain-flow counting method.

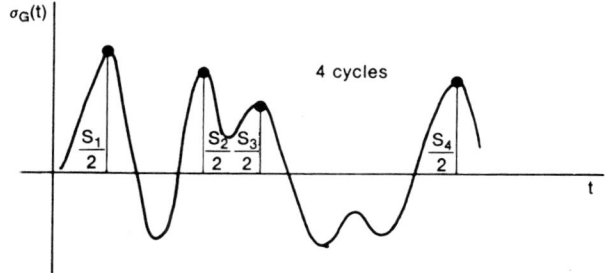

Fig. 8.8. Peak counting method.

### Range counting method (c)

A cycle period is defined by the time between two passages through zero with positive slope. The corresponding stress range S is the difference between the maximum and minimum geometric stress during this time interval (Fig. 8.9).

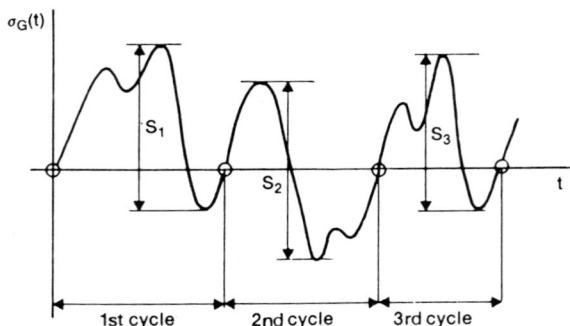

Fig. 8.9. Range counting method.

### Choice of a counting method (d)

Whatever the method adopted, all the cycles must be taken into account, including the smallest, because the S-N curve defined in Chapters 5 and 6 does not exhibit an endurance limit.

When the number of passages through the mean value is practically equal to the number of extremes, methods (b) and (c) are ideal for counting. If not (as in the case of many local extremes), it is better to use method (a).

## REFERENCES

8.1 Miner, M.A., Cumulative damage in fatigue, Transactions of the ASME, Vol.67, 1945.

8.2 Miner, M.A., Estimating fatigue life with particular emphasis on cumulative damage, in Metal Fatigue, G. Sines and J.L. Warsman editors, McGraw-Hill, 1959.

8.3 Palmgren, A., Die Lebensdauer von Kugellagern, Zitschrift des Vereins Deutscher Ingenieure, 68 (14), 1924.

8.4 Wirsching, P.H., and Yao, J.T.P., A probabilistic design approach using the Palmgren-Miner hypothesis, Methods of Structural Analysis, 1976.

8.5 Schutz, W., Fatigue life prediction, Mémoires et Etudes Scientifiques, Revue de Métallurgie, December 1982.

8.6 Bui-Quoc, T., Cumul de dommage en fatigue, Fatigue des matériaux et des structures, Compiègne University Collection, Editions Maloine.

8.7 Lemaître, J. and Chaboche, J.L., Aspect phenoménologique de la rupture par endommagement, Journal de Mécanique Appliquée, 2 (3), 1978.

8.8 Wirsching, P.H. and Mohsen Shehata, A., Fatigue under wide band random stresses using the rain flow method, Journal of Engineering Materials and Technology, July 1977, pp. 205-211.

8.9 Soize, C., Cumul de fatigue sous sollicitations aléatoires, Construction Métallique, No.4, 1979.

8.10 Dowling, N.E., Fatigue failure predictions for complicated stress/strain histories, Journal of Materials, JMLSA, 7 (1), March 1972.

8.11 Wirsching, P.H. and Light, M.C., Fatigue under wide band random stresses, Journal of the Structural Division, ST7, July 1980.

8.12 Strating, J., Fatigue and stochastic loadings, Thesis, Delft, 1973.

8.13 Bendat, J.S. and Piersol, A.G., Measurement and Analysis of Random Data, John Wiley and Sons, 1962.

# CHAPTER 9

# Fatigue Life Calculation
# by Fracture Mechanics

## 9.1 GENERAL

As shown below, in the description of the hypotheses underlying the use of a crack propagation law and the parameters of this rule, it is difficult at the present time, both from the experimental and theoretical standpoint, to consider the tools of fracture mechanics as a substitute for the standard approach of determining the service life of a tubular joint subject to fatigue.

Nevertheless, the so-called "fracture mechanics" approach which, subject to specific conditions and assumptions, estimates the service life by the integration of a crack propagation law, may prove to be an invaluable analytical tool in certain cases, especially when:

(a) The size of the initial defect, considered as a real crack, is known.

(b) An attempt is made to assess the influence on fatigue life of the variation in different design parameters (joint geometry, wall thickness, etc.).

(c) It is necessary to determine the sensitivity of fatigue life to the size of a defect.

*The evaluation of the fatigue service life by fracture mechanics consists of determining the number of cycles associated with a depth (or a form) of a given crack. For some specific applications, it may also be necessary to*

determine the size of the "initial defect"[1] connected
with a given number of cycles to "failure".

If the crack reaches a critical depth (or form),
characterized by a predetermined failure criterion, the
number of cycles calculated and associated with this
critical crack dimension is called the "number of cycles
to failure" or "fatigue life".

As a rule, the "critical" crack depth is considered to
be the depth defined by the conventional criterion of the
through crack (see Chapter 5). In a tubular joint this is
the crack that crosses the thickness of a tube wall.
Naturally, this is a conventional failure criterion, but
it does not necessarily correspond to the collapse of the
structure (Fig. 9.1).

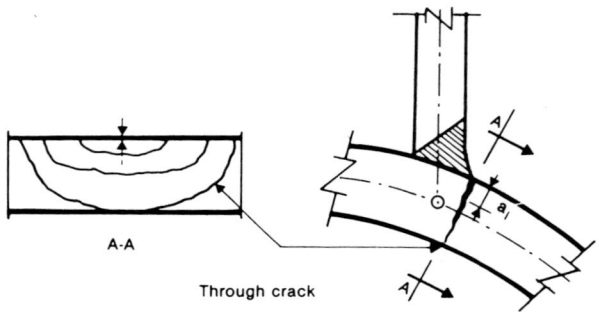

Fig. 9.1.

_____

(1) The term "initial defect" is placed in quotation marks
    to clearly distinguish the highly simplified image of
    this defect, equating the physical image of the
    initial defect distribution at the weld toe.

## 9.2  CRACK PROPAGATION LAW AND PROCEDURE FOR CALCULATING FATIGUE LIFE

The crack propagation law generally adopted is the following (Paris Law):

$$\frac{da}{dN} = C(\Delta K - \Delta K_S)^m$$

where

$\Delta K$ = the stress intensity factor range (sif) which depends on the load range applied, on the global joint geometry, on the local geometry at the junction and the crack geometry,

$\Delta K_S$ = "threshold" of $\Delta K$"[2] below which no significant crack propagation is observed,

a     = crack depth,

N     = the numer of cycles,

da/dN is the cracks propagation rate,

C and m are two constants depending on the material in which the crack is propagating.

As a rule, the crack propagates in the welded zone, namely in the heat affected zone (HAZ) and the deposited metal zone.

In the welded zone, the values of C and m are considerably scattered. In addition, the factors affecting C and m are not known accurately (residual stresses, load frequency, corrosive environment). In the absence of specific fixed or known values, it is recommended to take the following mean values:

m = 3.6 (this value of m results from tests on small specimens),

$$C = \frac{2.52 \cdot 10^{-5}}{(67)^m} \quad \text{(units daN, mm).}$$

---

(2) The term "threshold of $\Delta K$" is placed in quotation marks to point out the conventional connotation of $\Delta K_S$ to be introduced in a valid analytical form both for initiation and for propagation. It is also important to point out that in a simplified calculation, $\Delta K_S$ no longer has any relevance, due to the existence of a real crack at the weld toe.

These values result from a statistical analysis [9.1] and from the observations of several authors concerning the dependence of C and m.

The scatter in the values of $\Delta K_s$ is even greater than that of the coefficients m and C.

By integration of the Paris crack law, the service life is obtained by:

$$N_R = N_I + \frac{1}{C} \int_{a_i}^{a_f} \frac{da}{(\Delta K - \Delta K_s)^m}$$

where

$N_I$ = fatigue life for initiation of the crack of size $a_i$,

$N_R$ = fatigue life relative to conventional failure of a joint by the through crack criterion.

$a_f$ then corresponds to the thickness of the tube in which the initial defect $a_i$ has been detected.

*In the preliminary stages, the crack occurs in the form of many local incipient cracks, and then of a continuous front with a wide extension at the surface (over 100 mm) and shallow depth (a few millimetres).*

*This form is obviously not semi-elliptical any more than the forms in the subsequent phases. Furthermore, the angle of attack made by the crack front to the free surface is far from being a right angle, and inflexions have often been observed in the crack front due to local effects (Fig. 9.2).*

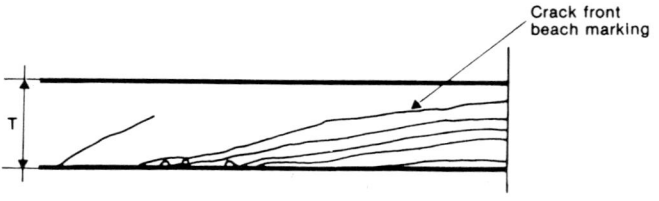

*Fig. 9.2.*

*The number of cycles to failure of a joint can be calculated simply using the Paris law, if the stress intensity factors are known for the geometry of the joint*

*and the loads acting on the crack plane, as well as the different correction factors to be taken into account.*

*On the whole, the crack propagation law takes account of the different parameters acting on fatigue life. This law is written:*

$$\frac{da}{dN} = f(G, C_{RS}, M, Q)$$

*where*

*G represents:*

*(a) The global geometry of the joint.*

*(b) The local geometry at the joint, profile and transition (weld connecting the brace to the chord).*

*(c) The "very local" geometry of the crack (its direction in relation to the principal stresses, its form, etc.).*

*$C_{RS}$ represents:*

*(a) The applied stress range.*

*(b) The ratio $R = \sigma_{min}/\sigma_{max}$.*

*M represents the influence of the environment (temperature, corrosion, etc.).*

*Q represents the characteristics of the base metal (non-propagation threshold, yield strength, ultimate strength, elongation at rupture, etc.).*

*The propagation law defined in the recommendations remains debatable especially since in reality, the crack propagates in the welded zone that is characterized by considerable heterogeneity, and where the residual welding stresses are complex.*

*However, specific analyses carried out using this law have shown that the law, albeit imperfect, yields results that show good agreement with test results.*

*The estimation of fatigue life by fracture mechanics normally requires the existence of a real initial crack with depth $a_i$ in the joint. This assumption amounts to ignoring the number of cycles related to crack initiation.*

*Different parameters may be involved in the estimation of $N_R$ by this model, but if only the parameter T (chord thickness) is isolated, the following equation can be confirmed [9.5, 9.6]:*

$$\frac{N_{R1}}{NR_2} = \left(\frac{T_2}{T_1}\right)^{m/2-1}$$

*This is only valid for:*

*(a) $\frac{a_i}{t}$ constant.*

*(b) Geometric parameters α, β, γ and τ constant.*

*(c) $\Delta K_S = 0$.*

*(d) $\dfrac{maximum\ principal\ stress\ on\ the\ outer\ surface}{maximum\ principal\ stress\ on\ the\ inner\ surface} = -1$.*

*Research is under way to investigate the influence of specific weld improvement treatments on fatigue life (see Chapter 7), and to determine the laws governing the initiation of a crack. The results of these investigations are likely to lead to new developments in the rules for fatigue analysis of welded joints, and towards a broader use of fracture mechanics.*

## 9.3  DETAILED ANALYSIS BY A NUMERICAL MODEL

### General presentation

As a part of the French research program on fatigue processes in welded tubular joints of offshore steel structures, a theoretical model to forecast the fatigue life taking account of the stress redistribution during cracking has been developed. It allows a constructive and quantitative consideration of the different causes (geometric, mechanical) which may influence the fatigue behavior of welded joints. One cause is the scale effect whose manifestations are complex, but in which the size parameter certainly plays an essential role.

Equally important are the indications provided by the model about the behavior of joints subject to large numbers of cycles. The crack propagation threshold then becomes a key parameter [9.2] , etc.).

### Remarks:

It should be noted that the implementation of this calculation demands a choice of assumptions, some of which depend on the structure investigated. In particular, in addition to the parameters involved in the basic assumptions, it is necessary to identify the propagation law and the local value of the factor $\Delta K$. This introduces a number of physical parameters whose determination is certainly not unique.

The calculated fatigue life depends on the choices and assumptions made. The calculation results cannot be examined without a critical analysis of these choices and assumptions.

#### General remarks:

This model was initially intended to enable the interpretation of the results obtained in the ECSC's "Offshore Technology" programme, and in particular, on the basis of fracture mechanics, to relate the results obtained in the experiments on test specimens with the fatigue behavior of large welded tubular structures.

More recently, an approach [9.7] using this model was adopted to provide a number of indications of the risks of brittle fracture in tubular joints. The stress intensity factor gradually tends towards a substantially uniform distribution along the crack front, and its value does not *exceed that which it had at the moment that the visible crack was detected.*

### Description of the model

*The proposed model was developped using the following basic assumptions:*

*1. It is considered that the crack propagates in each cross-section perpendicular to the weld toe in plane deformation conditions, at a rate depending on the local value of the stress intensity factor range $\Delta K$.*

*2. The random succession of initial defects is represented by a continuous pre-existing crack, with depth $a_O$, "invisible" to the observer.*

*3. At each point $\Psi$ , this crack propagates initially under the effect of the shell stress $\sigma_G(\Psi)$ multiplied by the local concentration factor $K_L$; this represents the crack initiation period.*

*4. The crack is declared to be "visible" at a point when it reaches a given depth $a_1 > a_O$; this serves to describe the crack propagation at the surface.*

*5. The propagation strictly speaking is modelled as follows. When the crack becomes "visible" in a joint, the stress abruptly drops to level $\sigma_G$: the local effect is nullified.*

*The stress redistribution then causes a gradual decrease in the shell stress proportional to the depth a reached; on the other hand, in the downstream sections, the stress increases at the approach of the visible crack.*

*6. The fatigue life of the joint corresponds to the time when the crack becomes a through crack at a given point.*

### Use of the numerical model

*For a given joint, this model can have two complementary uses. For the joint, it assumed that the geometry, the values characterizing the stress concentrations in the neighborhood of the weld, and the* characteristics in terms of the initial defect depth and the crack declared to be visible, are known. The parameters defining the stress redistribution in time are presumed to be substantially similar from one joint to another, and are hence "frozen" in the computer program. On the other hand, the user can adjust the duration of the initiation phase by adjusting the local effect parameter.

The primary use serves to predict the form of the crack as a function of the number of cycles, by using the crack length at the surface and the crack depth at any point of the weld toe. This numerical model quantitatively reflects the observations according to which the propagation of the crack at the surface in the period immediately following initiation is extremely rapid initially, and then slows down. Simultaneously, and by contrast, in depth propagation is very slow at the outset, and then gradually increases to reach a range that is stabilized or only rises slightly up to the through crack.

Another possibility consists of focusing attention exclusively on the central zone opposite the hot spot, and calculating the initiation time and number of cycles corresponding to the through crack. This is carried out for different loading levels selected automatically by the program, and hence allows the point-by-point plot of an S-N curve, in which S denotes the stress at the hot spot and N the number of cycles to failure, treated here as a through crack. Naturally, the second alternative, like the first, benefits from the representation - in terms of the local effect and the stress attenuation curve - of the stress redistribution. The local effect concerns the initiation phase and indicates a reorganisation on a limited scale, that of the weld, as opposed to the actual redistribution, which deals with the transfer, at the scale of the joint, of the forces towards the sections whose rigidity has not yet been affected significantly.

## REFERENCES

9.1    Recho, N. and Brozzetti, J., Design fatigue life of welded cruciform joints, IABSE Colloquium, Fatigue of Steel and Concrete Sructures, Lausanne, 1982.

9.2    Putot, C. and Radenkovic, D., Un modèle de propagation de fissures dans un noeud tubulaire soudé, Laboratoire de Mécanique des Solides, ANMT, Technical Report No.18, November 1980.

9.3    Putot, C., Notice d'utilisation du programme FATAL, IFP Report No.29687, November 1981.

9.4    Putot, C. and Frelat, J., Calcul de durée de vie des noeuds tubulaires du programme ANDF, ANMT Technical Report No.20, Septembre 1982.

9.5    Recho, N. and Brozzetti, J., Approche simplifiée du calcul de la durée de vie des noeuds tubulaires par la mécanique de la rupture, CTICM Report No.10.002.4, March 1982.

9.6    Recho, N. and Brozzetti, J., Influence de l'épaisseur de la membrure sur la durée de vie des noeuds tubulaires en T, Comparaison de deux approches, CTICM Note, October 1982.

9.7    Amiot, P., Radenkovic, D., Sanz, G. and Willm, P., Life prediction for tubular joints in offshore structures, 4th International Symposium of the Japan Welding Society, November 1982.

# ANNEXES

# NOTICE CONCERNING
# ANNEXES A AND B


## Correspondence between French and foreign steels
## for heavy metal plate and for steel tubes

### (Part I, Chapter 3)


These annexes are devoted to the problems of equivalences between foreign and french steel grades. This question is of the greatest concern to users and arises more and more frequently.

The growth of international trade, and the determination of French industry to intensify its export efforts, are causing manufacturers to work according to foreign drawings, requirements, standards and codes, with naturally define foreign steel grades.

Two alternativess are therefore available:

(1) Adopt the foreign grade specified, incurring the risk of procurement difficulties.
(2) Attempt to use a French grade, raising the problem of equivalence.


However, the readers's attention is drawn to the fact that this correspondence based on guaranteed values of fracture toughness is only approximate and given for information. For the same strength ranges, in fact, the guaranteed yield strengths may vary. Similarly, for the same quality indexes, the purities may be different. Morevover, particularly in the case of ASTM Standards, correspondence may differ according to the shape of the product and may also vary with the parameter selected: chemical composition and especially carbon content, yield strength and fracture toughness. Standards also do not always indicate guarantees against brittle fracture (guaranteed fracture toughness). Besides, the specimen sampling conditions for the analysis of mechanical properties are not identical in all the Standards, and this affects the guaranteed properties.

It is therefore vitally important for the user to refer always to the original Standard and to consult the steel companies or competent organizations for further details.

Before deciding on a grade, especially if a small quantity is involved, it is recommended to find out about the availability of the product and the minimum quantities.

# Correspondence
# between French and Foreign Steels
# for Heavy Metal Plate

## PURPOSE OF THE ANNEXE

This annexe gives the correspondences for:

(a) Type E24 steels with guaranteed fracture toughness at -40, -20, 0 and +20°C.

(b) Type E28 steels with guaranteed fracture toughness at -40, -20, 0 and +20°C.

(c) Type E36 steels with guaranteed fracture toughness at -40, -20, 0 and +20°C.

(d) Type E420 steels with guaranteed fracture toughness at -40, -20 and 0°C.

It also indicates:

(a) Thickness ranges specified by different Standards.

(b) Specific inspection conditions set by these Standards.

The properties given are those appearing in the Standards in force on 31 December 1982.

The ASTM grades considered are those appearing in the Specification API RP 2A.

TYPE E24 STEELS WITH GUARANTEED FRACTURE TOUGHNESS AT −40°C

| Standard | e (mm) | $C_{max.}$ | Residuals (%) | (*) $Re_{min.}$ ($N/mm^2$) | R ($N/mm^2$) | KV(L) (J) min. | KV(T) (J) min. |
|---|---|---|---|---|---|---|---|
| o BV |  |  |  |  |  |  |  |
| Grade E | ⩽ 50 | 0.18 |  | 235 | 400 − 490 | 27 |  |
|  | > 50 | 0.18 |  | 225/215/205 | 390 − 480 | 27 |  |
|  |  |  |  |  | 370 − 470 |  |  |
| o NF A 36 205 A 37 FP | ⩽ 110 | 0.16 | Cr ⩽ 0.25 Mo ⩽ 0.07 Ni ⩽ 0.30 Cu ⩽ 0.30 | 235/215/ 205/195 | 360 − 430 | 27 | 16 |
| o BS 4360 gr E |  | 0.16 |  | 260/245/ 240/225 | 400 − 480 | 27 at −50°C |  |
| o ASTM A 131 gr CS | | 0.16 | | 235 | 400 − 490 | 27[*] | 19[*] |
| gr E | | 0.18 | | 235 | 400 − 490 | 27 | 19 |

[*] for e ⩾ 50 mm.

*Values variable according to the thickness ranges specified by the standards.

TYPE E24 STEELS WITH GUARANTEED FRACTURE TOUGHNESS AT −20°C

| Standard | e (mm) | $C_{max.}$ | Residuals (%) | (*) $Re_{min.}$ $(N/mm^2)$ | R $(N/mm^2)$ | KV(L) (J) min. | KV(T) (J) min. |
|---|---|---|---|---|---|---|---|
| . NF A 35-501 E 24-4 | ⩽ 150 | 0.16 | | 235/215/205/ 195/185 | 340-460 | 27 | |
| . BV Grade D | ⩽ 50 | 0.21 | | 235 | 400-490 | 47at0°C (1) | |
| | > 50 | 0.21 | | 225/215/205 | 390-480 370-470 | 47at0°C (1) | |
| . NF A 36-205 A 37 FP AP | ⩽ 110 | 0.16 | Cr⩽0.25 Mo<0.07 Ni<0.30 Cu⩽0.30 | 235/215/205/ 195 | 360-430 | 32 | 21 |
| . DIN 17100 St 37-3 N | ⩽ 100 | 0.17 | | 235/225/215 | 340-470 | 27 | |
| . BS 4360 gr 40 D | | 0.16 (2) | | 260/245/240/ 225 | 400-480 | 27 | |
| gr 40 E | | 0.16 | | 260/245/240/ 225 | 400-480 | 61 | |

(1) See remark 3, Section 3.8.1, Part I, and remark on page "Type E24 steels with guaranteed fracture toughness at 0°C".

(2) Nb = 0.003 - 0.10      V = 0.003 - 0.10

*Values variable according to the thickness ranges specified by the standards.

TYPE E24 STEELS WITH GUARANTEED FRACTURE TOUGHNESS AT 0°C

| Standard | e (mm) | $C_{max.}$ | Residuals (%) | (*) $Re_{min.}$ $(N/mm^2)$ | R $(N/mm^2)$ | KV(L) (J) min. | KV(T) (J) min. |
|---|---|---|---|---|---|---|---|
| . NF A 35-501 E 24-3 | ⩽ 150 | 0.16 | | 235/215/205/ 195/185 | 340-460 | 27 | |
| . BV (1) Grade B | ⩽ 50 > 50 | 0.21 0.21 | | 235 225/215/205 | 400-490 390-480 370-470 | 27 27 | |
| . NF A 36-205 A 37 AP CP | ⩽110 | 0.16 | Cr ⩽ 0.25 Mo ⩽ 0.07 Ni ⩽ 0.30 Cu ⩽ 0.30 | 235/215/ 205/195 | 360-430 | 32 27 | 21 16 |
| . DIN 17100 St 37-3 U | ⩽100 | 0.17 | | 235/225/215 | 340-470 | 27 | |
| . BS 4360 gr 40 C | ⩽100 | 0.18 | | 230/225/ 220/210 | 400-480 | 27 | |
| . ASTM A 131 gr B gr D | | 0.21 0.21 | | 235 235 | 400-490 400-490 | 27 27 at -10°C | 19 19 at -10°C |
| A 516 gr 65 | ⩽ 13 13 to 50 50 to 100 | 0.24 0.26 0.28 | | 240 | 450-585 | } (2) | |
| A 573 gr 65 | ⩽ 13 13 to 38 | 0.24 0.26 | | 240 | 450-530 | } (2) | |
| A 709 gr 36 T 2 | ⩽ 19 19 to 38 38 to 64 64 to 100 | 0,25 0,25 0,26 0,27 | | 250 | 400-550 | 20 | |

(1) For Bureau Veritas grade D, a fracture toughness of 47J is guaranteed at 0°C, which is comparable with a grade with a guaranteed fracture toughness at -20°C.

(2) Specific requirements.

*Values variable according to the thickness ranges specified by the standards.

TYPE E24 STEELS WITH GUARANTEED FRACTURE TOUGHNESS AT +20°C

| Standard | e (mm) | $C_{max.}$ | Residuals (%) | (*) $Re_{min.}$ $(N/mm^2)$ | • $R$ $(N/mm^2)$ | KV(L) (J) min. | KV(T) (J) min. |
|---|---|---|---|---|---|---|---|
| . NF A 35-501 E 24-2 NE | ⩽ 30 30 to 150 | 0.17 0.19 | | 235 215/205/ .195/185 | 340-460 340-460 | 27 27 | |
| . BV Grade A | ⩽ 50 | C ⩽ 2.5 Mn | | 235 | 400-490 | (1) | |
| . NF A 36-205 A 37 CP | ⩽ 110 | 0.16 | Cr ⩽ 0.25 Mo ⩽ 0.07 Ni ⩽ 0.30 Cu ⩽ 0.30 | 235/215/ 205/195 | 360-430 | 32 | 21 |
| . DIN 17100 R St 37-2 | ⩽ 40 40 to 100 | 0.17 0.20 | | 235/225 215 | 340-470 340-470 | 27 27 | |
| . BS 4360 gr 40 B | ⩽ 100 | 0.20 | | 230/225/ 220/210 | 400-480 | 27 | |
| . ASTM A 36 | ⩽ 19 19 to 38 38 to 64 64 to 100 | 0.25 0.25 0.26 0.27 | | } 250 | } 400-550 | | |

(1) See remark 1, Section 3.8.1, Part I.

*Values variable according to the thickness ranges specified by the standards.

TYPE E28 STEELS WITH GUARANTEED FRACTURE TOUGHNESS AT −40°C

| Standard | e (mm) | $C_{max.}$ | Residuals (%) | (*) $Re_{min.}$ $(N/mm^2)$ | R $(N/mm^2)$ | KV(L) (J) min. | KV(T) (J) min. |
|---|---|---|---|---|---|---|---|
| . NF A 35-501 E 28-4 | ⩽ 150 | 0.18 | | 275/255/245/ 235/225 | 400-540 | 16 | |
| . NF A 36-205 A 47 FP | ⩽ 110 | 0.20 | Cr ⩽ 0.25 Mo ⩽ 0.10 Ni ⩽ 0.30 Cu ⩽ 0.30 | 295/275/265 | 470-540 | 40 | 21 |
| . BS 4360 gr 43 E | ⩽ 110 | 0.16 | | 280/270 / 255/240 | 430-510 | 27 at −50° C | |
| . ASTM A 633 gr A | ⩽ 110 | 0.18 (1) | | 290 | 430-570 | 34 (2) | 27 (2) |

(1) Nb ⩽ 0.05

(2) Only after agreement when ordering.

*Values variable according to the thickness ranges specified by the standards.

TYPE E28 STEELS WITH GUARANTEED FRACTURE TOUGHNESS AT −20°C

| Standard | e (mm) | $C_{max.}$ | Residuals (%) | (*) $Re_{min.}$ $(N/mm^2)$ | $R$ $(N/mm^2)$ | KV(L) (J) min. | KV(T) (J) min. |
|---|---|---|---|---|---|---|---|
| . NF A 35-501 E 28-4 | ≤ 150 | 0.18 | | 275/255/245/ 235/225 | 400-540 | 27 | |
| . NF A 36-205 A 48 FP AP | ≤ 110 | 0.20 | Cr ≤ 0,25 Mo ≤ 0.10 Ni ≤ 0.30 Cu ≤ 0.30 | 295/275/265 | 470-540 | 48 40 | 27 21 |
| . DIN 17100 St 44-3 R | ≤ 100 | 0,20 | | 275/265/255/ 245/235 | 410-540 | 27 | |
| . BS 4360 gr 43 D | ≤ 100 | 0.16 (1) | | 280/270/ 255/240 | 430-510 | 27 | |
| gr 43 E | ≤ 100 | 0.16 | | 280/270/ 255/240 | 430-510 | 61 | |
| . ASTM A 633 gr A | ≤ 100 | 0,18 (2) | | 290 | 430-570 | 54 (3) | 41 (3) |

(1) Nb = 0.003-0.10       V = 0.003-0.10

(2) Nb < 0.05

(3) Only after agreement when ordering.

*Values variable according to the thickness ranges specified by the standards.

TYPE E28 STEELS WITH GUARANTEED FRACTURE TOUGHNESS AT 0°C

| Standard | e (mm) | $C_{max.}$ | Residuals (%) | (*) $Re_{min.}$ $(N/mm^2)$ | R $(N/mm^2)$ | KV(L) (J) min. | KV(T) (J) min. |
|---|---|---|---|---|---|---|---|
| . NF A 35-501 E 28-3 | ⩽ 150 | 0.18 | | 275/255/245/ 235/225 | 400-480 | 27 | |
| . NF A 36-205 A 48 AP CP | ⩽ 110 | 0.20 | Cr ⩽ 0.25 Mo ⩽ 0.10 Ni ⩽ 0.30 Cu ⩽ 0.30 | 295/275/265 | 470-550 | 42 40 | 27 21 |
| . DIN 17100 St 44-3 -U | ⩽ 100 | 0.20 | | 275/265/255/ 245/235 | 410-540 | 27 | |

*Values variable according to the thickness ranges specified by the standards.

TYPE E28 STEELS WITH GUARANTEED FRACTURE TOUGHNESS AT +20°C

| Standard | e (mm) | $C_{max.}$ | Residuals (%) | (*) $Re_{min.}$ $(N/mm^2)$ | R $(N/mm^2)$ | KV(L) (J) min. | KV(T) (J) min. |
|---|---|---|---|---|---|---|---|
| . NF A 35-501 E 28-2 | ≤ 150 | 0.20 | | 275/255/245/ 235/225 | 400-540 | 27 | |
| . NF A 36-205 A 48 CP | ≤ 110 | 0.20 | Cr ≤ 0,25 Mo ≤ 0,10 Ni ≤ 0,30 Cu ≤ 0,30 | 295/275/265 | 470-550 | 42 | 27 |
| . DIN 17100 St 44-2 | ≤ 40 40 to 100 | 0.21 0.22 | | 275/265 255/245/235 | 410-540 410-540 | 27 27 | |
| . ASTM A 572 gr 42 | ≤ 150 | 0.21 | | 290 | ≥ 415 | | |

*Values variable according to the thickness ranges specified by the standards.

TYPE E36 STEELS WITH GUARANTEED FRACTURE TOUGHNESS AT −40°C

| Standard | e (mm) | $C_{max.}$ | Residuals (%) | (*) $Re_{min.}$ $(N/mm^2)$ | R $(N/mm^2)$ | KV(L) (J) min. | KV(T) (J) min. |
|---|---|---|---|---|---|---|---|
| . NF A 35-501 E 36-4 | ≤ 30 <br> 30 to 150 | 0.20 <br> 0.22 | | 355 <br> 325/315/305 | 490−630 <br> 490−630 | 27 <br> 27 | |
| . NF A 36-205 A 52 FP | ≤ 110 | 0.20 | Cr ≤ 0.25 <br> Mo ≤ 0.10 <br> Ni < 0.40 <br> Cu < 0.30 | 355/335/325 | 510−620 | 40 | 21 |
| . NF A 36-201 E 355 FP I | ≤ 35 <br> 35 to 100 | 0.18 (1) <br> 0.20 (1) | Cr ≤ 0.25 <br> Mo ≤ 0.10 <br> Ni < 0.30 <br> Cu ≤ 0.35 | 355 <br> 335/325 | 480−610 <br> 480−610 | 40 <br> 40 | 20 <br> 20 |
| E 355 FP II | ≤ 35 <br> 35 to 100 | 0.16 (2) <br> 0.18 (2) | idem | 355 <br> 335/325 | 480−610 <br> 480−610 | 40 <br> 40 | 20 <br> 20 |
| . BV Grade EH 36 | ≤ 50 <br> > 50 | 0.18 (3) <br> 0.20 (3) | Cr ≤ 0,20 <br> Mo ≤ 0,08 <br> Ni < 0,40 <br> Cu ≤ 0,35 | 355 <br> 335/325/315 | 490−620 <br> 480−610 <br> 470−600 | 34 <br> 34 | |
| . BS 4360 gr 50 E | ≤ 16 <br> 16 to 63 | 0.18 (4) <br> 0.20 (4) | | 355 <br> 345/340 | 490−620 <br> 490−620 | 41 at −35°C or 27 at −50°C | |
| . A.S.T.M. A 131 gr EH 36 | | 0.18 (5) | idem B.V. | 360 | 490−620 | 34 | 23 |
| A 633 gr C et D | ≤ 100 | 0.20 (6) | id B.V. (7) | 345/315 | 480−620 <br> 450−590 | 34 (5) | 27 (8) |

(1) Nb = 0.010−0.060
(2) Nb = 0.010−0.060     V = 0.020−0.10
(3) Nb = 0.020−0.050     V = 0.050−0.10
(4) Nb = 0.003−0.10     V = 0.003−0.15
(5) Nb ≤ 0.05     V ≤ 0.10
(6) gr C : Nb = 0,01−0,05
(7) Only gr D.
(8) Only after agreement when ordering.
*Values variable according to the thickness ranges specified by the standards.

TYPE E36 STEELS WITH GUARANTEED FRACTURE TOUGHNESS AT −20°C

| Standard | e (mm) | $C_{max.}$ | Residuals (%) | (*) $Re_{min.}$ (N/mm$^2$) | R (N/mm$^2$) | KV(L) (J) min. | KV(T) (J) min. |
|---|---|---|---|---|---|---|---|
| . NF A 35-501 E 36-4 | ⩽ 30 | 0,20 | | 355 | 490-630 | 42 | |
| | 30 to 150 | 0,22 | | 325/315/305 | 490-630 | 42 | |
| . NF A 36-205 A 52 FP | ⩽ 110 | 0,20 | Cr ⩽ 0,25 | 355/335/325 | 510-620 | 48 | 27 |
| AP | | | Mo ⩽ 0.10 | | | 40 | 21 |
| | | | Ni ⩽ 0,40 | | | | |
| | | | Cu ⩽ 0,30 | | | | |
| . NF A 36-201 E 355 R I | ⩽ 35 | 0,18 (1) | Cr ⩽ 0,25 | 355 | 480-610 | 40 | 20 |
| | 35 to 100 | 0,20 (1) | Mo ⩽ 0.10 | 335/325 | 480-610 | 40 | 20 |
| | | | Ni ⩽ 0,30 | | | | |
| | | | Cu ⩽ 0,35 | | | | |
| E 355 R II | ⩽ 35 | 0,16 (2) | } idem | 355 | 480-610 | 40 | 20 |
| | 35 to 100 | 0,18 (2) | | 335/325 | 480-610 | 40 | 20 |
| E 355 FP I | ⩽ 35 | 0,18 (1) | } idem | 355 | 480-610 | 48 | 27 |
| | 35 to 100 | 0,20 (1) | | 335/325 | 480-610 | 48 | 27 |
| E 355 FP II | ⩽ 35 | 0,16 (2) | } idem | 355 | 480-610 | 48 | 27 |
| | 35 to 100 | 0,18 (2) | | 335/325 | 480-610 | 48 | 27 |
| . BV Grade DH 36 | ⩽ 50 | 0.18 (3) | Cr ⩽ 0.20 | 355 | 490-620 | 34 | |
| | | | Mo ⩽ 0.08 | | | | |
| | | | Ni ⩽ 0,40 | | | | |
| | > 50 | 0.20 (3) | Cu ⩽ 0,35 | 335/325/315 | 480-610 | 34 | |
| | | | | | 470-600 | | |
| . DIN 17100 St 52-3 N | ⩽ 30 | 0,20 | | 355 | 490-630 | 27 | |
| | 30 to 100 | 0,22 | | 345/335/ 325/315 | 490-630 | 27 | |

.../...

*Values variable according to the thickness ranges specified by the standards.

TYPE E36 STEELS WITH GUARANTEED FRACTURE TOUGHNESS AT −20°C
(continued)

| Standard | e (mm) | C max. | Residuals (%) | (*) Re min. (N/mm$^2$) | R (N/mm$^2$) | KV(L) (J) min. | KV(T) (J) min. |
|---|---|---|---|---|---|---|---|
| . BS 4360 | | | | | | | |
| gr 50 D | ⩽ 16 | 0.18 (4) | | 355 | 490–620 | 41 | |
| | 16 to 40 | 0.20 (4) | | 345 | 490–620 | 41 | |
| gr 50 E | ⩽ 16 | 0.18 (4) | | . 355 | 490–620 | 47 at −30° C | |
| | 16 to 63 | 0.20 (4) | | 345/340 | 490/620 | 47 at −30° C | |
| . ASTM A 131 | | | | | | | |
| gr DH 36 | | 0.18 (5) | Cr ⩽ 0.25 Mo ⩽ 0.08 Ni ⩽ 0.40 Cu ⩽ 0.35 | 360 | 490–620 | 34 | 23 |
| A 537 cl 1 | ⩽ 38 | 0.24 | Cr ⩽ 0.25 Mo ⩽ 0.08 Ni ⩽ 0.25 | 345 | 485–620 | supple- mentary requi- rements | |
| | 38 to 100 | 0.24 | Cu ⩽ 0.35 | 310 | 450–585 | | |
| A 633 gr C and | ⩽ 100 | 0.20 (6) | idem (7) | 345/315 | 480–620 450–590 | 54 (8) | 41 (8) |

(1) Nb = 0.010-0,060
(2) Nb = 0.010=0.060    V = 0.02-0,10
(3) Nb = 0.020-0.050    V = 0.050-0.10
(4) Nb = 0.003-0.10     V = 0.003-0.10 (0,15 for gr 50 E)
(5) Nb ⩽ 0.05           V ⩽ 0,10
(6) gr C : Nb = 0.01-0.05
(7) Only gr D.                    (8) Only after agreement when ordering.
*Values variable according to the thickness ranges specified by the standards.

TYPE E36 STEELS WITH GUARANTEED FRACTURE TOUGHNESS AT 0°C

| Standard | e (mm) | $C_{max.}$ | Residuals (%) | (*) $Re_{min.}$ $(N/mm^2)$ | R $(N/mm^2)$ | KV(L) (J) min. | KV(T) (J) min. |
|---|---|---|---|---|---|---|---|
| . NF A 35-501 E 36-3 | < 16 | 0.20 | | 355 | 490-630 | 27 | |
| | 16 to 150 | 0.22 | | 335/315/305 | 490-630 | 27 | |
| . NF A 36-205 A 52 AP | < 110 | 0.20 | Cr < 0.25 | 355/335/315 | 510-620 | 42 | 27 |
| CP | | | Mo < 0.10 | | | 40 | 21 |
| | | | Ni < 0.30 | | | | |
| | | | Cu < 0.30 | | | | |
| . NF A 36-201 E 355 R·I | ≤ 35 | 0.18 (1) | Cr < 0.25 | 355 | 480-610 | 42 | 27 |
| | | | Mo < 0.10 | | | | |
| | | | Ni < 0.30 | | | | |
| | 35 to 100 | 0.20 (1) | Cu < 0.35 | 335/325 | 480-610 | 42 | 27 |
| E 355 R II | < 35 | 0.16 (2) | idem | 355 | 480-610 | 42 | 27 |
| | 35 to 100 | 0.18 (2) | | 335/325 | 480-610 | 42 | 27 |
| . BV Grade AH 36 | < 50 | 0.18 (3) | Cr ≤ 0.20 | 355 | 490-620 | 34 | |
| | | | Mo ≤ 0.08 | | | | |
| | | | Ni ≤ 0.40 | | | | |
| | | | Cu ≤ 0.35 | | | | |
| | > 50 | 0.20 (3) | idem | 335/325/315 | 480-610 470-600 | 34 | |

.../...

*Values variable according to the thickness ranges specified by the standards.

TYPE E36 STEELS WITH GUARANTEED FRACTURE TOUGHNESS AT 0°C
(continued)

| Standard | e (mm) | $C_{max.}$ | Residuals (%) | (*) $Re_{min.}$ ($N/mm^2$) | R ($N/mm^2$) | KV(L) (J) min. | KV(T) (J) min. |
|---|---|---|---|---|---|---|---|
| . DIN 17100 St 52-3 U | ≤ 30 | 0.20 | | 355 | 490-630 | 27 | |
| | 30 to 100 | 0.22 | | 345/335/ 325/315 | 490-690 | 27 | |
| . BS 4360 gr 50 C | ≤ 16 | 0.20 (4) | | 355 | 490-630 | 41 at -5° C | |
| | 16 to 100 | 0.22 (4) | | 345/340/325 | 490-630 | 41 at -5° C | |
| . ASTM A 131 gr AH 36 | | 0.18 (3) | Cr ≤ 0.25 Mo ≤ 0.08 Ni ≤ 0.40 Cu ≤ 0.35 | 360 | 490-620 | 34 | 23 |
| A 709 gr 50 T 2 | ≤ 50 | 0.23 | | 345 | > 450 | 20 | |
| gr 50 T 3 | ≤ 50 | 0.23 | | 345 | > 450 | 20 at - 10°C | |

(1) Nb = 0.010-0.060
(2) Nb = 0.010-0.060    V = 0.02-0.10
(3) Nb ≤ 0.05    V ≤ 0.10
(4) Nb = 0.003-0.10    V = 0.003-0.10

*Values variable according to the thickness ranges specified by the standards.

TYPE E36 STEELS WITH GUARANTEED FRACTURE TOUGHNESS AT +20°C

| Standard | e (mm) | $C_{max.}$ | Residuals (%) | (*) $Re_{min.}$ $(N/mm^2)$ | R $(N/mm^2)$ | KV(L) (J) min. | KV(T) (J) min. |
|---|---|---|---|---|---|---|---|
| . NF A 36-205 A 52 CP | $\leqslant$ 110 | 0.20 | Cr $\leqslant$ 0.25 Mo $\leqslant$ 0.10 Ni $\leqslant$ 0.30 Cu $\leqslant$ 0.30 | 355/335/325 | 510-620 | 42 | 27 |
| . BS 4360 gr 50 B | $\leqslant$ 16 16 to 100 | 0.20 (1) 0.22 (1) | | 355 345/340/325 | 490-620 490-620 | | |
| . ASTM A 441 A 572 gr 50 | $\leqslant$ 100 $\leqslant$ 50 | 0.22 (2) 0.23 | | 345/315/290 345 | $\geqslant$ 485/460 $\geqslant$ 450 | | |

(1) Nb = 0.003 - 0.10          V = 0.003 - 0.10

(2) V $\geqslant$ 0.02

*Values variable according to the thickness ranges specified by the standards.

TYPE E420 STEELS WITH GUARANTEED FRACTURE TOUGHNESS AT −40°C

| Standard | e (mm) | $C_{max.}$ | Residuals (%) | (*) $Re_{min.}$ $(N/mm^2)$ | R $(N/mm^2)$ | KV(L) (J) min. | KV(T) (J) min. |
|---|---|---|---|---|---|---|---|
| . NF A 36-201 E 420 I FP | ⩽ 80 | 0.20 (1) | Cr ⩽ 0.25 Mo ⩽ 0.10 Ni ⩽ 0.30 Cu ⩽ 0.35 | 420/410/400 | 530-670 | 40 | 20 |
| E 420 II FP | ⩽ 80 | 0.22 | Cr ⩽ 0.40 Mo ⩽ 0.40 Ni ⩽ 0.70 Cu ⩽ 0.60 | 420/410/400 | 530-670 | 40 | 20 |
| . BS 4360 gr 55 E | ⩽100 | 0.22 (2) | | 450/430/ 415/410 | 550-700 | 27 at -50°C | |
| . ASTM A 633 gr E | ⩽100 | 0.22 (3) | | 415 | 550-690 | 34 (4) | 27 (4) |

(1) Nb = 0.010 - 0.060     V = 0.02 - 0.12
(2) Nb = 0.003 - 0.10     V = 0.003 - 0.20
(3) Nb ⩽ 0.05
(4) Only after agreement when ordering.

*Values variable according to the thickness ranges specified by the standards.

TYPE E420 STEELS WITH GUARANTEED FRACTURE TOUGHNESS AT −20°C

| Standard | e (mm) | $C_{max.}$ | Residuals (%) | (*) $Re_{min.}$ ($N/mm^2$) | R ($N/mm^2$) | KV(L) (J) min. | KV(T) (J) min. |
|---|---|---|---|---|---|---|---|
| .NF A 36-201 E 420 I R | ≤ 80 | 0.20 (1) | Cr ≤ 0.25 Mo ≤ 0.10 Ni ≤ 0.30 Cu ≤ 0.35 | 420/410/400 | 530-670 | 40 | 20 |
| E 420 II R | ≤ 80 | 0.22 | Cr ≤ 0.40 Mo ≤ 0.40 Ni ≤ 0.70 Cu ≤ 0.60 | 420/410/400 | 530-670 | 40 | 20 |
| E 420 I FP | ≤ 80 | 0.20 (1) | Cr ≤ 0.25 Mo ≤ 0.10 Ni ≤ 0.30 Cu ≤ 0.35 | 420/410/400 | 530-670 | 42 | 27 |
| E 420 II FP | ≤ 80 | 0.22 | Cr ≤ 0.40 Mo ≤ 0.40 Ni ≤ 0.70 Cu ≤ 0.60 | 420/410/400 | 530-670 | 42 | 27 |
| . BS 4360 gr 55 E | ≤ 100 | 0.22(2) | | 450/430/ 415/400 | 550-700 | 61 | |
| . ASTM A 633 gr E | ≤ 100 | 0.22(3) | | 415 | 550-690. | 54 (4) | 41 (4) |

(1) Nb = 0.010-0.060     V = 0.02-0.12
(2) Nb = 0.003-0.10      V = 0.003-0.20
(3) Nb ≤ 0.05
(4) Only after agreement when ordering.·

*Values variable according to the thickness ranges specified by the standards.

TYPE E420 STEELS WITH GUARANTEED FRACTURE TOUGHNESS AT 0°C

| Standard | e<br>(mm) | $C_{max.}$ | Residuals<br>(%) | (*)<br>$Re_{min.}$<br>$(N/mm^2)$ | R<br>$(N/mm^2)$ | KV(L)<br>(J)<br>min. | KV(T)<br>(J)<br>min. |
|---|---|---|---|---|---|---|---|
| . NF A 36-201<br>  E 420 I R | ≤ 80 | 0.20(1) | Cr ≤ 0.25<br>Mo ≤ 0.10<br>Ni ≤ 0.30<br>Cu ≤ 0.35 | 420/410/400 | 530-670 | 42 | 27 |
| E 420 II R | ≤ 80 | 0.22 | Cr ≤ 0.40<br>Mo ≤ 0.40<br>Ni ≤ 0.70<br>Cu ≤ 0.60 | 420/410/400 | 530-670 | 42 | 27 |
| . BS 4360<br>  gr 55 C | ≤ 63 | 0.22(2) | | 450/430/415 | 550-700 | 27 | |

(1) Nb = 0.010-0.060      V = 0.02-0.12

(2) Nb = 0.003-0.10      V = 0.003-0.20

*Values variable according to the thickness ranges specified by the standards.

CARBON STEELS DEFINED BY FRENCH STANDARDS FOR TUBES
SUITABLE FOR USE FOR THE CONSTRUCTION OF OFFSHORE PLATFORMS

TABLE 1. Steels with $R_{p\,0.2}$ min. guaranteed $< 300$ N/mm$^2$

| Test temperature (°C) | Values of $R_{p\,0.2}$ (N/mm$^2$) | Guaranteed fracture toughness (J/cm$^2$) | | | | Grades | Reference Standards |
|---|---|---|---|---|---|---|---|
| | | Lengthwise | | Transverse | | | |
| | | mean | ind | mean | ind | | |
| − 45°C | 220 235 275 | 35 " " | 28 " " | | | TU 37 b₃ " 42 " " 48 " | NF A 49-211 TSS Ø 10.2 to 610 |
| | 290 | 40 | 32 | | | TU E 290 b₃ | NF A 49-411 TSS Ø 60.3 to 406.4 |
| | 240 | 35 | 28 | | | TU 42 BT | NF A 49-230 TSS Ø 13.5 to 406.4 |
| | 240 | 35 | 28 | | | TS 42 BT | NF A 49-240 TS-ERW Ø 13.5 to 406.4 |

Note :

(1) NF A 49-410, which offers the same guaranteed fracture
    thoughness as NF A 49-211 will probably be withdrawn.
    It is therefore deliberately absent from this table.

(2) TSS = seamless tubes.
    TS = welded tubes.

(3) The figures appearing after the standards correspond to the
    date of the latest edition

CARBON STEELS DEFINED BY FRENCH STANDARDS FOR TUBES
SUITABLE FOR USE FOR THE CONSTRUCTION OF OFFSHORE PLATFORMS

TABLE 1. Steels with $R_{p\,0.2}$ min. guaranteed $< 300$ N/mm$^2$

| Test temperature (°C) | Values of $R_{p\,0.2}$ (N/mm$^2$) | Guaranteed fracture toughness (J/cm$^2$) | | | | Grades | Reference Standards |
|---|---|---|---|---|---|---|---|
| | | Lengthwise | | Transverse | | | |
| | | mean | ind | mean | ind | | |
| − 20° C | 220 | 35 | 28 | | | TU 37 b$_2$ | NF A 49-211 |
| | 235 | " | " | | | " 42 " | |
| | 275 | " | " | | | " 48 " | |
| | 290 | 40 | 32 | | | TU E 290 b$_2$ | NF A 49-411 |
| | 235 | 35 | 26 | | | TS or TU E24.4 | NF A 49-501 |
| | 255 | 35 | 26 | | | " " " E26.4 | TS hot finished |
| | 295 | 50 | 37 | | | " " " E30.4 | and TSS |
| | | | | | | | ⃝21.3-1220 |
| | | | | | | | ▢90 - 350 |
| | | | | | | | ▭50.25-400. 300 |
| | 235 | 35 | 26 | | | TS E 24.4 | NF A 49-541 |
| | 275 | 35 | 26 | | | TS E 28.4 | TS cold finished |
| | | | | | | | ⃝21.3-1620 |
| | | | | | | | ▢22 - 140 |
| | | | | | | | ▭35.20-180. 100 |

CARBON STEELS DEFINED BY FRENCH STANDARDS FOR TUBES
SUITABLE FOR USE FOR THE CONSTRUCTION OF OFFSHORE PLATFORMS

TABLE 1. Steels with $R_{p\,0.2}$ min. guaranteed $< 300$ N/mm$^2$

| Test temperature (°C) | Values of $R_{p\,0.2}$ (N/mm$^2$) | Guaranteed fracture toughness (J/cm$^2$) | | | | Grades | Reference Standards |
| | | Lengthwise | | Transverse | | | |
| | | mean | ind | mean | ind | | |
|---|---|---|---|---|---|---|---|
| 0°C | 220<br>235<br>275 | 35<br>"<br>" | 28<br>"<br>" | | | TU 37 b$_1$<br>"  42 b$_1$<br>"  48 b$_1$ | NF A 49-211 |
| | 295 | 35 | 28 | | | TU E 290 b$_1$ | NF A 49-411 |
| | 235<br>255<br>295 | 35<br>"<br>" | 26<br>"<br>" | | | TS or TU E 24.3<br>"  "  "  E 26.3<br>"  "  "  E 30.3 | NF A 49-501 |
| | 235<br>275 | 35<br>" | 26<br>" | | | TS E 24.3<br>"  E 28.3 | NF A 49-541 |
| | 220<br>250<br>290 | 70<br>"<br>" | 50<br>"<br>" | 35<br>"<br>" | 28<br>"<br>" | TS E 220<br>"  E 250<br>"  E 290 | NF A 49-400-TS<br>ERW<br>Ø 17,2-406,4 |
| | 220<br>250<br>290 | | | 35<br>"<br>" | 28<br>"<br>" | TS E 220 b<br>"  E 250 b<br>"  E 290 b | NFA 49-401-TS.<br>SAW<br>Ø 406,4-1220 |
| | 225<br>245<br>285 | 35<br>35<br>50 | | 20<br>24<br>26 | | TS 37 CP<br>"  42 CP<br>"  48 CP | NF A 49-252 - TS<br>use  :+≤350°C<br>ERW.Ø168,3-406,4<br>SAW.Ø168,3-1220 |
| | 225<br>245<br>285 | 35<br>35<br>50 | 26<br>26<br>38 | 20<br>24<br>26 | 15<br>18<br>20 | TS 37 CP<br>"  42 CP<br>"  48 CP | NF A 49-253-TS<br>use  ·+>350°C<br>SAW:Ø 457-1220 |
| | 220<br>235<br>275 | 40<br>40<br>50 | 30<br>30<br>35 | | | TU 37 c<br>"  42 c<br>"  48 c | NF A 49-213<br>TSS Ø 17,2-610 |

CARBON STEELS DEFINED BY FRENCH STANDARDS FOR TUBES
SUITABLE FOR USE FOR THE CONSTRUCTION OF OFFSHORE PLATFORMS

TABLE 1. Steels with $R_{p\,0.2}$ min. guaranteed $< 300$ N/mm²

| Test temperature (°C) | Values of $R_{p\,0.2}$ (N/mm²) | Guaranteed fracture toughness (J/cm²) | | | | Grades | Reference Standards |
|---|---|---|---|---|---|---|---|
| | | Lengthwise | | Transverse | | | |
| | | mean | ind | mean | ind | | |
| + 20°C | 225 | 40 | | 26 | | TS 37 CP | NF A 49-252 |
| | 245 | 40 | | 30 | | "   42 " | |
| | 285 | 60 | | 35 | | "   48 " | |
| | 225 | 40 | 30 | 26 | 20 | TS 37 CP | NF A 49-253 |
| | 245 | 40 | 30 | 30 | 24 | "   42   " | |
| | 285 | 60 | 45 | 35 | 26 | "   48   " | |
| | 235 | 60 | 45 | | | TS or TU E24.2 | NF A 49-501 |
| | 255 | 60 | 45 | | | "  "  "  E26.2 | |
| | 295 | 60 | 45 | | | "  "  "  E30.2 | |
| | 235 | 35 | 26 | | | TS E 24.2 | NF A 49-541 |
| | 275 | 35 | 26 | | | "  E 28.2 | |

CARBON STEELS DEFINED BY FRENCH STANDARDS FOR TUBES
SUITABLE FOR USE FOR THE CONSTRUCTION OF OFFSHORE PLATFORMS

TABLE 2. Steels with 300 N/mm$^2$ $\leqslant R_{p\,0.2}$ min. guaranteed $\leqslant$ 420 N/mm$^2$

| Test temperature (°C) | Values of $R_{p\,0.2}$ (N/mm$^2$) | Guaranteed fracture toughness (J/cm$^2$) | | | | Grades | Reference Standards |
|---|---|---|---|---|---|---|---|
| | | Lengthwise | | Transverse | | | |
| | | mean | ind | mean | ind | | |
| − 45°C | 320 | 40 | 32 | | | TUE 320 b$_3$ | NF A 49-411 |
| | 360 | 40 | 32 | | | TUE 360 " | NF A 49-411 |
| | 415 | 40 | 32 | | | TUE 415 " | NF A 49-411 |
| − 20°C | 320 | 40 | 32 | | | TUE 320 b$_2$ | NF A 49-411 |
| | 360 | 40 | 32 | | | " 360 " | " " |
| | 415 | 40 | 32 | | | " 415 " | " " |
| | 355 | 50 | 37 | | | TUandTS E36.4 | NF A 49-501 |
| 0°C | 320 | 35 | 28 | | | TUE 320 b$_1$ | NF A 49-411 |
| | 360 | " | " | | | 360 "$^1$ | |
| | 415 | " | " | | | 415 " | |
| | 355 | 35 | 26 | | | TUandTS E36.3 | NF A 49-501 |
| | 320 | 70 | 50 | 35 | 28 | TS E 320 | NF A 49-400 |
| | 360 | 70 | 50 | 40 | 32 | TS E 360 | |
| | 415 | 70 | 50 | 40 | 32 | TS E 415 | |
| | 320 | | | 35 | 28 | TS E 320 b | NF A 49-401 |
| | 360 | | | 40 | 32 | TS E 360 b | |
| | 335 | 50 | | 26 | | TS 52 CP | NF A 49-252 |
| | 335 | 50 | 38 | 26 | 20 | TS 52 CP | NF A 49-253 |
| + 20°C | 335 | 60 | | 35 | | TS 52 CP | NF A 49-252 |
| | 335 | 60 | 45 | 35 | 26 | TS 52 CP | NF A 49-253 |
| | 355 | 60 | 45 | | | TU or TS E36.2 | NF A 49-501 |

ANNEXE **B**

# French Standards
# for Steel Tubes

TYPE E24 STEELS WITH GUARANTEED FRACTURE TOUGHNESS AT $-45°C$

| Reference standard | Diameter D (mm) | $C_{max}$ (as cast)* | Residuals (%) | $R_{p\,0.2}$ min. (N/mm$^2$) | $R_m$ (N/mm$^2$) | Mean KCV (J/cm$^2$) Lengthwise | Mean KCV (J/cm$^2$) Transverse |
|---|---|---|---|---|---|---|---|
| NF A 49-211 - TSS - | 10.2 (1.6-2.3) 610 (10 - 40) | | | | | | |
| TU 37 b₃ | | 0.18 | | 220 | 360-480 | 35 | |
| TU 42 b₃ | | 0.22 | | 235 | 410-530 | 35 | |
| NF A 49-230 - TSS - | 13.5 x 2.3 406.4 (8.8 à 30) | | | | | | |
| TU 42 BT | | 0.20 | | 240 | 415-510 | 35 | |
| NF A 49-240 - TS - | 13.5 (2-2.3) 406.4 (4.5-10) | | | | | | |
| TS 42 BT | | 0.20 | | 240 | 415-510 | 35 | |
| Pr DIN 17 174 - TS - | DIN 2458 | | | | | (at -40°C) | (at -40°C) |
| TT St 35 N(1.0356) | e ≤ 25 | 0.17 | | 225 | 340-460 | 40 | 27 |
| TT St 35 V(" " ") | e ≤ 40 | 0.17 | | 235 | 360-490 | 40 | 27 |
| Pr DIN 17 173 - TSS - | DIN 2448 | | | | | (at -40°C) | |
| TT St 35 N(1.0356) | e ≤ 25 | 0.17 | | 225 | 340-460 | 40 | 27 |
| TT St 35 V(" " ") | e ≤ 40 | 0.17 | | 235 | 360-490 | 40 | 27 |
| BS 3600 | BS 3600 | | | | | (at -50°C) Charpy V long) | |
| 410 {SS / S | | 0.20 | | 235 | 410-530 | 27 | |
| ASTM A 333 Grade 6 | ANSI B 16.10 | 0.30 | | 241 | 414 | 17.6 | |

*The as cast value was adopted as the one most commonly given in foreign standards. Add two points on the average for the product.

TYPE E24 STEELS WITH GUARANTEED FRACTURE TOUGHNESS AT −20°C

| Reference standard | Diameter D (mm) | $C_{max}$ (as cast)* | Residuals (%) | $R_{p\,0.2}$ min. (N/mm$^2$) | $R_m$ (N/mm$^2$) | Mean KCV (J/cm$^2$) Lengthwise | Transverse |
|---|---|---|---|---|---|---|---|
| NF A 49-211 − TSS − <br> TU 37 b$_2$ <br> TU 42 b$_2$ | 10.2 (1.6−2.3) <br> 610 (10−40) | 0.18 <br> 0.22 | | 220 <br> 235 | 360−480 <br> 410−530 | 35 <br> 35 | |
| NF A 49-501 <br> TS } E 24.4 <br> or TU } | O 21.3 x 2.3 to 1220 (10−40) <br> □ 22 x 2.3 to 350 ( 8−12) <br> ▭ 50 x 25 x 2.6 to 400 − 300 (8−12) | TU 0.18 <br> TS 0.16 | | 235 | 320−460 | 35 | |
| NF A 49-541 <br> TS E 24.4 | O 21.3 x 2.3 to 1620 (10−16) <br> □ 22 x 2.3 to 140 ( 4−5 ) <br> ▭ 35 x 20 x 2.5 to 180 x 100 x 3 | | | 235 | 360 | 35 | |
| ISO R 630 <br> Fe 360 D | Hollow sections according to ISO R 657 | 0.17 | | 235 | 360−460 | 27 | |

TYPE E24 STEELS WITH GUARANTEED FRACTURE TOUGHNESS AT 0°C

| Reference standard | Diameter D (mm) | $C_{max}$ (as cast)* | Residuals (%) | $R_p$ 0.2 min. (N/mm$^2$) | $R_m$ (N/mm$^2$) | Mean KCV (J/cm$^2$) Lengthwise | Transverse |
|---|---|---|---|---|---|---|---|
| **NF A 49-211** | | | | | | | |
| TU 37 b] | 10.2 (1.6-2.3) | 0.18 | | 220 | 360-480 | 35 | |
| TU 42 b] | to 610 (10 - 40) | 0.22 | | 235 | 410-530 | 35 | |
| **NF A 49-501** | O 21.3 x 2.3 to 1220 (10-40) | | | | | | |
| TS E 24.3 or TU | □ 22 x 2,3 to 350 (8 - 12) | TU 0.18 | | 235 | 320-460 | 35 | |
| | ▭ 50 x 25 x 2.6 to 400 - 300 (8-12) | TS 0.16 | | | | | |
| **NF A 49-541** | O 21.3 x 2.3 to 1620 (10-16) | | | | | | |
| TS E 24.3 | □ 22 x 2,3 (4 - 5) to 140 | 0.16 | | 235 | 360 | 35 | |
| | ▭ 35 x 20 x 2.5 to 180 x 100 x 3 | | | | | | |
| **NF A 49-401** | 406.4 (5 - 16) | | | | | | |
| TS E 220 b | to 1220 (6.3-36) | 0.16 | | 220 | 370-480 | 35 | 35 |
| TS E 250 b | | 0.20 | | 250 | 420-530 | 35 | 35 |
| **NF A 49-400** | 17.2 x 2.3 | | | | | | |
| TS E 220 | to 406.4 (6,3-8,8) | 0.15 | | 220 | 370-490 | 70 | 35 |
| TS E 250 | | 0.16 | | 250 | 410-530 | 70 | 35 |

TYPE E24 STEELS WITH GUARANTEED FRACTURE TOUGHNESS AT 0°C

| Reference standard | Diameter D (mm) | $C_{max}$ (as cast)* | Residuals (%) | $R_{p\,0.2}$ min. (N/mm²) | $R_m$ (N/mm²) | Mean KCV (J/cm²) Lengthwise | Mean KCV (J/cm²) Transverse |
|---|---|---|---|---|---|---|---|
| NF A 49-252 | 168.3 (4 - 7,1) to 1220 (10 -50) | | | | | | |
| TS 37 CP | | 0.16 | Cu + 6Sn ≤ 0,33 | 225 | 360-480 | 35 | 20 |
| TS 42 CP | | 0.18 | Cu ≤ 0,18 | 245 | 410-520 | 35 | 24 |
| NF A 49-253 | 457 (5-12,5) to 1220 (10-50) | | | | | | |
| TS 37 CP | | 0.16 | Cu + 6Sn ≤ 0,33 | 225 | 360-480 | 35 | 20 |
| TS 42 CP | | 0.18 | Cu ≤ 0,18 | 245 | 410-530 | 35 | 24 |
| NF A 49-213 | 17.2 (2,3-4) to 610 (10-50) | | | | | | |
| TU 37 c | | 0.16 | Cu ≤ 0,25 | 220 | 360-460 | 40 | |
| TU 42 c | | 0.20 | Sn ≤ 0,030 | 235 | 410-510 | 40 | |
| DIN 17 172 | DIN 2448 DIN 2458 | | | | | ≤ 500mm | > 500 mm |
| TS - TSS / St E 240.7 | | 0.22 | | 240 | 370-490 | 47 | 27 |
| API 5 L (83 edition) | Additional fracture toughness requirements (for information) | | | | | | |
| ISO R 630 | Hollow sections ISO R 657 | | | | | | |
| Fe 360 C | | 0.17 | | 235 | 430-530 | 27 | |

TYPE E24 STEELS WITH GUARANTEED FRACTURE TOUGHNESS AT +20°C

| Reference standard | Diameter D (mm) | $C_{max}$ (as cast)* | Residuals (%) | $R_{p\,0.2}$ min. (N/mm²) | $R_m$ (N/mm²) | Mean KCV (J/cm²) Lengthwise | Mean KCV (J/cm²) Transverse |
|---|---|---|---|---|---|---|---|
| NF A 49-252 | | | | | | | |
| TS 37 CP | 168.3 (4-7,1) to 1220 (10-50) | 0.16 | Cu + 6Sn ⩽ 0.33 | 225 | 360-480 | 40 | 26 |
| TS 42 CP | | 0.18 | Cu ⩽ 0.18 | 245 | 410-520 | 40 | 30 |
| NF A 49-253 | | | | | | | |
| TS 37 CP | 457 (5-12,5) to 1220 (10-50) | 0.16 | Cu + 6Sn ⩽ 0,33 | 225 | 360-480 | 40 | 26 |
| TS 42 CP | | 0.18 | Cu ⩽ 0.18 | 245 | 410-530 | 40 | 30 |
| NF A 49-541 | | | | | | | |
| TS E 24.2 | O 21.3 x 2.3 to 1620 (10-16) □ 22 x 2.3 to 140 (4-5) ▭ 35 x 20 x 2.5 to 180 x 100 x 3 | 0.17 | | 235 | 360 | 35 | |
| NF A 49-501 | | | | | | | |
| TU or TS E 24.2 | O 21.3 x 2.3 to 1220 (10-40) □ 22 x 2.3 to 350 (8-12) ▭ 50 x 25 x 2.6 to 400 x 300 (8-12) | TU ou TS 0.20 | | 235 | 320-460 | 60 | |
| DIN 1629/4 -TSS- (Pr DIN 1630) | | | | | | | |
| St 37.4(1.0255) | DIN 2448 | 0.17 | | 235 | 350-480 | 43 | 27 |
| DIN 1626/4 -TS- (Pr DIN 1628) | | | | | | | |
| St 37.4(1.0255) | DIN 2458 | 0.17 | | 235 | 350-480 | 43 | 27 |

TYPE E24 STEELS WITH GUARANTEED FRACTURE TOUGHNESS AT +20°C

| Reference standard | Diameter D (mm) | $C_{max}$ (as cast)* | Residuals (%) | $R_{p\,0.2}$ min. (N/mm²) | $R_m$ (N/mm²) | Mean KCV (J/cm²) Lengthwise | Transverse |
|---|---|---|---|---|---|---|---|
| DIN 17 175 -TSS- St 35.8(1.0305) | | 0.17 | | 235 | 360-480 | D V M | 34 |
| BS 3601 TS and BS 3602 TSS 410 | | BS 3601 0.17 BS 3602 0.20 | | 245 | 410-550 | KCV to be agreed with users | |
| ISO R 630 Fe 360 B Hollow sections ISO R 657/XIV | | 0.18 | | 235 | 360-460 | 27 | |

TYPE E28 STEELS WITH GUARANTEED FRACTURE TOUGHNESS AT −45°C

| Reference standard | Diameter D (mm) | $C_{max}$ (as cast)* | Residuals (%) | $R_{p\,0.2}$ min. $(N/mm^2)$ | $R_m$ $(N/mm^2)$ | Mean KCV $(J/cm^2)$ Lengthwise | Transverse |
|---|---|---|---|---|---|---|---|
| NF A 49-211<br>TU 48 $b_3$ | 10.2 (1,6 - 2,3)<br>610 (10 - 40) | 0,23 | | 275 | 470-590 | 35 | |
| NF A 49-411<br>TU E 290 $b_3$ | 60.3 (2,9 - 11)<br>406.4 (8,8 - 40) | 0,21 | | 290 | 420-540 | 40 | |
| Pr DIN 17 178 -TS- | DIN 2458 | | | | | ( t = − 40°C) | |
| T St E 255 (1.0463) | | 0,16 | Cr+Cu+Mo ⩽0,45 | 255 | 360-480 | 31 | 20 |
| T St E 285 (1.0488) | | 0,16 | | 285 | 390-510 | 31 | 20 |
| Pr DIN 17 179 -TSS- | DIN 2448 | | | | | (t = −40°C) | |
| T St E 255 (1.0463) | | 0,16 | Cr+Cu+Mo ⩽0,45 | 255 | 360-480 | 31 | 20 |
| T St E 285 (1.0488) | | 0,16 | | 285 | 390-510 | 31 | 20 |

TYPE E28 STEELS WITH GUARANTEED FRACTURE TOUGHNESS AT −20°C

| Reference standard | Diameter D (mm) | $C_{max}$ (as cast)* | Residuals (%) | $R_{p\,0.2}$ min. (N/mm²) | $R_m$ (N/mm²) | Mean KCV (J/cm²) Lengthwise | Transverse |
|---|---|---|---|---|---|---|---|
| NF A 49-211<br>TU E 48 b₂ | 10.2 (1,6 - 2,3)<br>610 (10 - 40) | 0.23 | | 275 | 470-590 | 35 | |
| NF A 49-411<br>TU E 290 b₂ | 60.3 (2,9 - 11)<br>406.4 (8,8 - 40) | 0.21 | | 290 | 420-540 | 40 | |
| NF A 49-501<br>or<br>TS ou TU E 26.4<br>TS ou TU E 30.4<br>or | O 21.3 x 2.3<br>to 1220 (10-40)<br>□ 22 x 2.3<br>to 350 ( 8-12)<br>□ 50 x 25 x 2.6<br>to 400 - 300 (8-12) | TU 0.22<br>TS 0.18<br>TU 0.22<br>TS 0.20 | | 255<br>295 | 370-510<br>410-550 | 35<br>50 | |
| NF A 49-541<br>TS E 28.4 | O 21.3 x 2.3<br>to 1620 (10-16)<br>□ 22 x 2.3<br>to 140 ( 4- 5)<br>□ 35 x 20 x 2.5<br>to 180 x 100 x 3 | 0.18 | | 275 | 400 | 35 | |
| Pr DIN 17 178<br>-TS-<br>T St E 255 (1.0463)<br>T St E 285 (1.0488) | DIN 2458 | 0.16<br>0.16 | Cr+Cu+Mo ≤ 0.45 | 255<br>285 | 360-480<br>390-510 | 47<br>47 | 27<br>27 |
| Pr DIN 17 179<br>-TSS-<br>T St E 255 (1.0463)<br>T St E 285 (1.0488) | DIN 2448 | 0.16<br>0.16 | Cr+Cu+Mo ≤ 0.45 | 255<br>285 | 360-480<br>390-510 | 47<br>27 | 47<br>27 |
| ISO R 630<br>Fe 430 B | Hollow sections<br>ISO R 657 | 0.21 | | 275 | 430-530 | 27 | |

TYPE E28 STEELS WITH GUARANTEED FRACTURE TOUGHNESS AT 0°C

| Reference standard | Diameter D (mm) | $C_{max}$ (as cast)* | Residuals (%) | $R_{p\,0.2}$ min. (N/mm²) | $R_m$ (N/mm²) | Mean KCV (J/cm²) | |
|---|---|---|---|---|---|---|---|
| | | | | | | Lengthwise | Transverse |
| NF A 49-211<br>TU 48 b₁ | 10.2 (1.6-2.3)<br>610 (10 - 40) | 0.23 | | 275 | 470-590 | 35 | |
| NF A 49-411<br>TU E 290 b₁ | 60.3 (2.9-11)<br>406.4 (8.8-40) | 0.21 | | 290 | 420-540 | 35 | |
| NF A 49-501<br>TS or TU E 26.3<br>TS or TU E 30.3 | O 21.3 x 2.3<br>to 1220 (10-40)<br>□ 22 x 2.3<br>to 350 ( 8-12)<br>▭ 50 x 25 x 2.6<br>to 400 - 300 (8-12) | TS 0,18<br>TU 0,22<br>TS 0,20<br>TU 0,22 | | 255<br>295 | 370-510<br>410-550 | 35 | |
| NF A 49-541<br>TS E 28.3 | O 21.3 x 2.3<br>to 1620 (10-16)<br>□ 22 x 2.3<br>to 140 (4-5)<br>▭ 35 x 20 x 2.5<br>to 180 x 100 x 3 | 0.18 | | 275 | 400 | 35 | |
| NF A 49-400<br>TS E 290 | 17,2 x 2,3<br>406.4 (6.3-8.8) | 0.16 | | 290 | 420-540 | 70 | 35 |
| NF A 49-401<br>TS E 290 b | 406.4 (5-16)<br>to 1220 (6.3 - 36) | 0.20 | | 290 | 420-530 | | 35 |

TYPE E28 STEELS WITH GUARANTEED FRACTURE TOUGHNESS AT 0°C

| Reference standard | Diameter D (mm) | $C_{max}$ (as cast)* | Residuals (%) | $R_{p\,0.2}$ min. (N/mm²) | $R_m$ (N/mm²) | Mean KCV (J/cm²) Lengthwise | Mean KCV (J/cm²) Transverse |
|---|---|---|---|---|---|---|---|
| NF A 49-252 <br> TS 48 CP | 168,3 (4-7,1) <br> to 1220 (10-50) | 0,20 | Ni ≤ 0,30 <br> Cr ≤ 0,25 <br> Mo ≤ 0,10 | 285 | 470-590 | 50 | 26 |
| NF A 49-253 <br> TS 48 CP | 457 (5-12.5) <br> to 1220 (10-50) | 0,20 | Ni ≤ 0,30 <br> Cr ≤ 0,25 <br> Mo ≤ 0,10 | 285 | 470-590 | 50 | 26 |
| NF A 49-213 <br> TU 48 c | 17,2 (2,3-4) <br> to 610 (10-50) | 0,22 | Cu ≤ 0,25 <br> Sn ≤ 0,030 | 275 | 470-570 | 50 | |
| Pr DIN 17 178 -TS- <br> T St E 255 (1.0463) <br> T St E 285 (1.0488) | DIN 2458 | 0,16 <br> 0,16 | Cr+Cu+Mo ≤ 0.45 | 255 <br> 285 | 360-480 <br> 390-510 | (t = -40°C) <br> 55 <br> 55 | (t = -40°C) <br> 31 <br> 31 |
| Pr DIN 17 179 -TSS- <br> T St E 255 (1.0463) <br> T St E 285 (1.0488) | DIN 2448 | 0,16 <br> 0,16 | Cr+Cu+Mo ≤ 0.45 | 255 <br> 285 | 360-480 <br> 390-510 | (t = -40°C) <br> 55 <br> 55 | (t = -40°C) <br> 31 <br> 31 |
| DIN 17 172 <br> St E 290.7(1.0484) | TSS DIN 2448 <br> TS DIN 2458 | 0,22 | | 290 | 420-540 | ∅ ≤ 500mm <br> 47 | ∅ > 500mm <br> 27 |
| API 5 L (83 edition) | Fracture toughness: additional requirements (for information) | | | | | | |
| ISO R 630 <br> Fe 430 C | ISO/ R 657/XIV | 0,20 | | 275 | 430-530 | 27 | |

TYPE E28 STEELS WITH GUARANTEED FRACTURE TOUGHNESS AT +20°C

| Reference standard | Diameter D (mm) | $C_{max}$ (as cast)* | Residuals (%) | $R_{p\,0.2}$ min. (N/mm²) | $R_m$ (N/mm²) | Mean KCV (J/cm²) Lengthwise | Transverse |
|---|---|---|---|---|---|---|---|
| NF A 49-252 <br> TS 48 CP | 168.3 ( 4- 7.1) to 1220 (10-50) | 0,20 | Ni ≤ 0.30 <br> Cr ≤ 0.25 <br> Mo ≤ 0.10 | 285 | 470-590 | 60 | |
| NF A 49-253 <br> TS 48 CP | 457 ( 5-12.5) to 1220 (10-50) | 0,20 | Ni ≤ 0.30 <br> Cr ≤ 0.25 <br> Mo ≤ 0.10 | 285 | 470-590 | 60 | |
| NF A 49-541 <br><br> TS E 28.2 | ○ 21.3 x 2.3 to 1620 (10-16) <br> □ 22 x 2.3 to 140 ( 4- 5) <br> ▭ 35 x 20 x 2.5 to 180 x 100 x 3 | 0.21 | | 275 | 400 | 35 | |
| NF A 49-501 <br><br> TS or TU E 26.2 <br> TS or TU E 30.2 | ○ 21.3 x 2.3 to 1220 (10-40) <br> □ 22 x 2.3 to 350 ( 8-12) <br> ▭ 50 x 25 x 2.6 to 400 - 300 (8-12) | TS 0.23 <br> TU 0.25 <br> TS 0.25 <br> TU 0.25 | | 255 <br> 295 | 370-510 <br> 410-550 | 60 <br> 60 | |
| Pr DIN 17 178 -TS- <br><br> T St E 255 (1.0463) <br> T St E 285 (1.0488) | DIN 2458 | 0.16 <br> 0.16 | Cr+Cu+Mo ≤ 0.45 | 255 <br> 285 | 360-480 <br> 390-510 | 63 <br> 63 | 39 <br> 39 |

TYPE E28 STEELS WITH GUARANTEED FRACTURE TOUGHNESS AT +20°C

| Reference standard | Diameter D (mm) | $C_{max}$ (as cast)* | Residuals (%) | $R_{p\,0.2}$ min. $(N/mm^2)$ | $R_m$ $(N/mm^2)$ | Mean KCV $(J/cm^2)$ Lengthwise | Transverse |
|---|---|---|---|---|---|---|---|
| Pr DIN 17 179 -TSS- T St E 255 (1.0463) T St E 285 (1.0488) | DIN 2448 | 0.16 0.16 | Cr+Cu+Mo ≤ 0.45 | 255 285 | 360-480 390-510 | 63 63 | 39 39 |
| DIN 1629/4 -TSS- (Pr DIN 1630) St 44.4 (1.0257) | DIN 2448 | 0.20 | | 275 | 420-550 | 43 | 27 |
| DIN 1626/4 -TS- (Pr DIN 1628) St 44.4 (1.0257) | DIN 2458 | 0.20 | | 275 | 420-550 | 43 | 27 |
| DIN 17 175 -TSS- St 45.8 | DIN 2448 | 0.21 | | 255 | 410-530 | | DVM 27 |
| BS 3601 BS 3602 460 | | (BS 3601) 0.22 (BS 3602) 0.20 | | 280 | 460-600 | KCV to be agreed with users | |
| ISO R 630 Fe 430 B | Hollow sections ISO R 657 | 0.21 | | 275 | 430-530 | 27 | |

TYPE E36 STEELS WITH GUARANTEED FRACTURE TOUGHNESS AT –45°C

| Reference standard | Diameter D (mm) | $C_{max}$ (as cast)* | Residuals (%) | $R_{p\,0.2}$ min. (N/mm²) | $R_m$ (N/mm²) | Mean KCV (J/cm²) | |
|---|---|---|---|---|---|---|---|
| | | | | | | Lengthwise | Transverse |
| NF A 49-411 | 60.3 (2.9 - 11) 406.4 (8.8 - 40) | | | | | | |
| TU E 320 b₃ | | 0.21 | | 320 | 460–580 | 40 | |
| TU E 360 b₃ | | 0.21 | | 360 | 510–630 | 40 | |
| Pr DIN 17 178 -TS- T St E 355 (1.0566) | DIN 2458 | 0.18 | Cr+Cu+Mo ≤ 0,45 | 355 | 490–630 | ( – 40°C) 31 | 20 |
| Pr DIN 17 179 -TSS- T St E 355 (1.0566) | DIN 2448 | 0.18 | Cr+Cu+Mo ≤ 0,45 | 355 | 490–630 | ( – 40°C) 31 | 20 |

TYPE E36 STEELS WITH GUARANTEED FRACTURE TOUGHNESS AT −20°C

| Reference standard | Diameter D (mm) | $C_{max}$ (as cast)* | Residuals (%) | $R_{p\ 0.2}$ min. (N/mm²) | $R_m$ (N/mm²) | Mean KCV (J/cm²) Lengthwise | Transverse |
|---|---|---|---|---|---|---|---|
| NF A 49-411<br>TU E 320 b₂<br>TU E 360 b₂ | 60.3 (2.9 −11)<br>406.4 (8.8 −40) | 0.21<br>0.21 | | 320<br>360 | 460−580<br>510−630 | 40<br>40 | |
| NF A 49-501<br><br><br>TU or TS E 36.4 | ○ 21.3 x 2.3 to 1220 (10−40)<br>□ 22 x 2.3 to 350 (8 −12)<br>▭ 50 x 25 x 2.6 to 400 − 300 (8−12) | TU and TS<br>0,20 | | 355 | 470−610 | 50 | |
| Pr DIN 17 178<br>−TS−<br>T St E 355 (1.0566) | DIN 2458 | 0,18 | Cr+Cu+Mo ≤ 0,45 | 355 | 490−630 | 47 | 27 |
| Pr DIN 17 179<br>−TSS−<br>T St E 355 (1.0566) | DIN 2448 | 0,18 | Cr+Cu+Mo ≤ 0,45 | 355 | 490−630 | 47 | 27 |
| IŚO R 630<br>Fe 510 D | Hollow sections<br>ISO R 657/XIV | 0,20 | | 355 | 490−630 | 27 | 27 |

TYPE E36 STEELS WITH GUARANTEED FRACTURE TOUGHNESS AT 0°C

| Reference standard | Diameter D (mm) | $C_{max}$ (as cast)* | Residuals (%) | $R_{p\,0.2}$ min. (N/mm²) | $R_m$ (N/mm²) | Mean KCV (J/cm²) Lengthwise | Transverse |
|---|---|---|---|---|---|---|---|
| NF A 49-411<br>TU E 320 b₁<br>TU E 360 b₁ | 60.3 ( 2.9-11 )<br>406.4 ( 8.8-40 ) | 0.21<br>0.21 | | 320<br>360 | 460-580<br>510-630 | 35<br>35 | |
| NF A 49-501 | ○ 21.3 x 2.3 to 1220 (10 -40)<br>□ 22 x 2.3 to 350 (8 -12)<br>□ 50 x 25 x 2,6 to 400 - 300 (8-12) | TU and TS | | | | | |
| TS E 36.3 | | 0.20 | | 355 | 470-610 | 35 | |
| NF A 49-400<br>TS E 320<br>TS E 360 | 17.2 x 2.3 to 406.4 ( 6.3-8.8) | 0.16<br>0.16 | | 320<br>360 | 440-560<br>480-620 | 70<br>70 | 35<br>40 |
| NF A 49-401<br>TS E 320 b<br>TS E 360 b | 406.4 ( 5 -16) to 1220 ( 6.3-36) | 0.21<br>0.22 | | 320<br>360 | 440-560<br>480-620 | | 35<br>40 |
| NF A 49-252<br>TS 52 CP | 168.3 ( 4 -7.1) to 1220 (10 -50) | 0,20 | | 335 | 510-630 | 50 | 26 |
| NF A 49-253<br>TS 52 CP | 457 ( 5 -12.5) to 1220 (10 -50) | 0.20 | | 335 | 510-630 | 50 | 26 |

TYPE E36 STEELS WITH GUARANTEED FRACTURE TOUGHNESS AT 0°C

| Reference standard | Diameter D (mm) | $C_{max}$ (as cast)* | Residuals (%) | R p 0.2 min. (N/mm²) | $R_m$ (N/mm²) | Mean KCV (J/cm²) | |
|---|---|---|---|---|---|---|---|
| | | | | | | Lengthwise | Transverse |
| Pr DIN 17 178 -TS- T St E 355 (1.0566) | DIN 2458 | 0.18 | Cr+Cu+Mo ≤ 0.45 | 355 | 490-630 | 55 | 31 |
| Pr DIN 17 179 -TSS- T St E 355 (1.0566) | DIN 2448 | 0.18 | Cr+Cu+Mo ≤ 0.45 | 355 | 490-630 | 55 | 31 |
| DIN 17 172 TS and TSS St E 320.7 St E 360.7 St E 387.7 | DIN 2448 DIN 2458 | 0.22 0.22 0.22 | | 320 360 385 | 460-580 510-630 530-680 | ø ≤ 500mm 47 47 47 | ø > 500 mm 27 27 27 |
| API 5 L (83 edition) | Fracture toughness: additional requirements (for information) | | | | | | |
| ISO R 630 Fe 510 C | ISO R 657/XIV | 0.20 | | 355 | 490-630 | 27 | |

TYPE E36 STEELS WITH GUARANTEED FRACTURE TOUGHNESS AT +20°C

| Reference standard | Diameter D (mm) | $C_{max}$ (as cast)* | Residuals (%) | $R_{p\,0.2}$ min. (N/mm²) | $R_m$ (N/mm²) | Mean KCV (J/cm²) Lengthwise | Mean KCV (J/cm²) Transverse |
|---|---|---|---|---|---|---|---|
| NF A 49-252 -TS- TS 52 CP | 168,3 ( 4- 7,1) to 1220 (10-50) | 0.20 | | 335 | 510-630 | 60 | 35 |
| NF A 49-253 -TS- TS 52 CP | 457 ( 5-12,5) to 1220 (10-50) | 0.20 | | 335 | 510-630 | 60 | 35 |
| NF A 49-501 TU or TS E 36.2 | O 21,3 x 2,3 to 1220 (10-40) □ 22 x 2,3 to 350 ( 8-12) □ 50 x 25 x 2,6 to 400-300 (8-12) TU and TS | 0.25 | | 355 | 470-610 | 60 | |
| Pr DIN 17 178 -TS- T St E 355 (1.0566) | DIN 2458 | 0.18 | Cr+Cu+Mo ⩽ 0.45 | 355 | 490-630 | 63 | 39 |
| Pr DIN 17 179 -TSS- T St E 355 (1.0566) | DIN 2448 | 0.18 | Cr+Cu+Mo ⩽ 0.45 | 355 | 490-630 | 63 | 39 |
| DIN 1629/4 -TS- (Pr DIN 1630) St 52.4 (1.0581) | DIN 2448 | 0.22 | | 355 | 500-650 | 43 | 27 |
| DIN 1626/4 -TS- (Pr DIN 1628) St 52.4 | DIN 2458 | 0.22 | | 355 | 500-650 | 43 | 27 |
| ISO R 630 Fe 510 B | Hollow sections ISO R 657 | 0.22 | | 355 | 490-630 | 27 | |

TYPE E420 STEELS WITH GUARANTEED FRACTURE TOUGHNESS AT −45°C.

| Reference standard | Diameter D (mm) | $C_{max}$ (as cast)* | Residuals (%) | R p 0.2 min. (N/mm²) | $R_m$ (N/mm²) | Mean KCV (J/cm²) Lengthwise | Mean KCV (J/cm²) Transverse |
|---|---|---|---|---|---|---|---|
| NF A 49-411 TU E 415 b₃ | 60,3 (2.9-11) 406,4 (8.8-40) | 0.18 | | 415 | 550-700 | 40 | |
| Pr DIN 17 178 -TS- T St E 460 (1.8915) | DIN 2458 | 0.20 | Cr ≤ 0,30 Cu ≤ 0,20 Mo ≤ 0,10 | 460 | 560-730 | 31 (− 40°C) | 20 |
| Pr DIN 17 179 -TSS- T St E 460 (1.8915) | DIN 2448 | 0.20 | Cr ≤ 0,30 Cu ≤ 0,20 Mo ≤ 0,10 | 460 | 560-730 | 31 (− 40°C) | 20 |

TYPE E420 STEELS WITH GUARANTEED FRACTURE TOUGHNESS AT -20°C

| Reference standard | Diameter D (mm) | $C_{max}$ (as cast)* | Residuals (%) | $R_{p\,0.2}$ min. (N/mm²) | $R_m$ (N/mm²) | Mean KCV (J/cm²) | |
|---|---|---|---|---|---|---|---|
| | | | | | | Lengthwise | Transverse |
| NF A 49-411 TU E 415 b₂ | 60,3 (2,9 - 11) 406,4 (8,8 - 40) | 0,18 | | 415 | 550-700 | 40 | |
| Pr DIN 17 178 -TS- T St E 460 (1.8915) | DIN 2458 | 0,20 | Cr ≤ 0,30 Cu ≤ 0,20 Mo ≤ 0,10 | 460 | 560-730 | 47 | 27 |
| Pr DIN 17 179 -TSS- T St E 460 (1.8915) | DIN 2448 | 0,20 | Cr ≤ 0,30 Cu ≤ 0,20 Mo ≤ 0,10 | 460 | 560-730 | 47 | 27 |

TYPE E420 STEELS WITH GUARANTEED FRACTURE TOUGHNESS AT 0°C

| Reference standard | Diameter D (mm) | $C_{max}$ (as cast)* | Residuals (%) | $R_p$ 0.2 min. (N/mm²) | $R_m$ (N/mm²) | Mean KCV (J/cm²) | |
|---|---|---|---|---|---|---|---|
| | | | | | | Lengthwise | Transverse |
| NF A 49-411<br>TU E 415 $b_1$ | 60.3 (2,9–11)<br>406.4 (8,8–40) | 0.18 | | 415 | 550–700 | 35 | |
| NF A 49-400<br>TS E 415 | 17.2 x 2.3<br>to 406.4 (6,3–8,8) | 0.20 | | 415 | 530–680 | 70 | 40 |
| Pr DIN 17 178<br>-TS-<br>T St E 460 (1.8915) | DIN 2458 | 0.20 | $Cr \leqslant 0.30$<br>$Cu \leqslant 0.20$<br>$Mo \leqslant 0.10$ | 460 | 560–730 | 55 | 31 |
| Pr DIN 17 179<br>-TSS-<br>T St E 460 (1.8915) | DIN 2448 | 0.20 | $Cr \leqslant 0.30$<br>$Cu \leqslant 0.20$<br>$Mo \leqslant 0.10$ | 460 | 560–730 | 55 | 31 |
| DIN 17 172<br>TS-TSS<br>St E 415.7 | DIN 2448<br>DIN 2458 | 0.22 | | 415 | 550–700 | $\emptyset \leqslant 500mm$<br>47 | $\emptyset > 500$ mm<br>27 |
| API 5 L (83 edition) | | | | | | | |

Fracture toughness: additional requirements (for information)

TYPE E420 STEELS WITH GUARANTEED FRACTURE TOUGHNESS AT +20°C

| Reference standard | Diameter D (mm) | $C_{max}$ (as cast)* | Residuals (%) | $R_{p\ 0.2}$ min. $(N/mm^2)$ | $R_m$ $(N/mm^2)$ | Mean KCV $(J/cm^2)$ Lengthwise | Transverse |
|---|---|---|---|---|---|---|---|
| NF – Nil | | | | | | | |
| Pr DIN 17 178 –TS– T St E 460 (1.8915) | DIN 2458 | 0.20 | Cr ⩽ 0.30 Cu ⩽ 0.20 Mo ⩽ 0,10 | 460 | 560–730 | 63 | 39 |
| Pr DIN 17 179 –TSS– T St E 460 (1.8915) | DIN 2448 | 0.20 | Cr ⩽ 0.30 Cu ⩽ 0.20 Mo ⩽ 0.10 | 460 | 560–730 | 63 | 39 |

# Stress Concentration Factors in Tubular Joints

# Parametric Formulas
# EPR, DnV, Lloyd's

## DEFINITIONS AND NOTATIONS

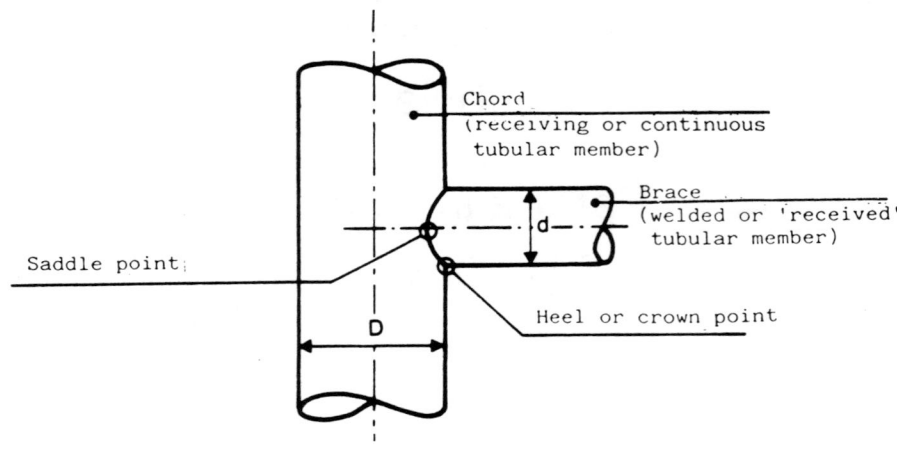

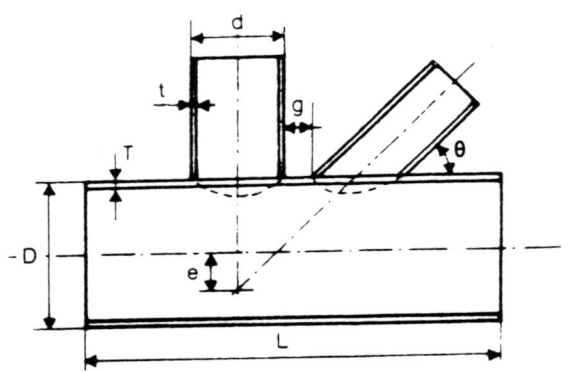

Fig. 1.2.

L = chord stub length,
D = chord outside diameter,
T = chord thickness,
d = brace outside diameter,
t = brace thickness,
g = theoretical gap,
e = eccentricity (positive in Fig. 1.2, negative otherwise),
Θ = acute angle defining the brace inclination,
α = 2L/D chord stub slenderness ratio,
β = d/D brace to chord diameter ratio,
γ = D/2T parameter defining the slenderness of the chord wall,
τ = t/T brace thickness to chord thickness ratio,
ζ = g/D relative gap.

In the case of two or more braces, they are identiried by a subscript.

## CLASSIFICATION OF TUBULAR JOINTS

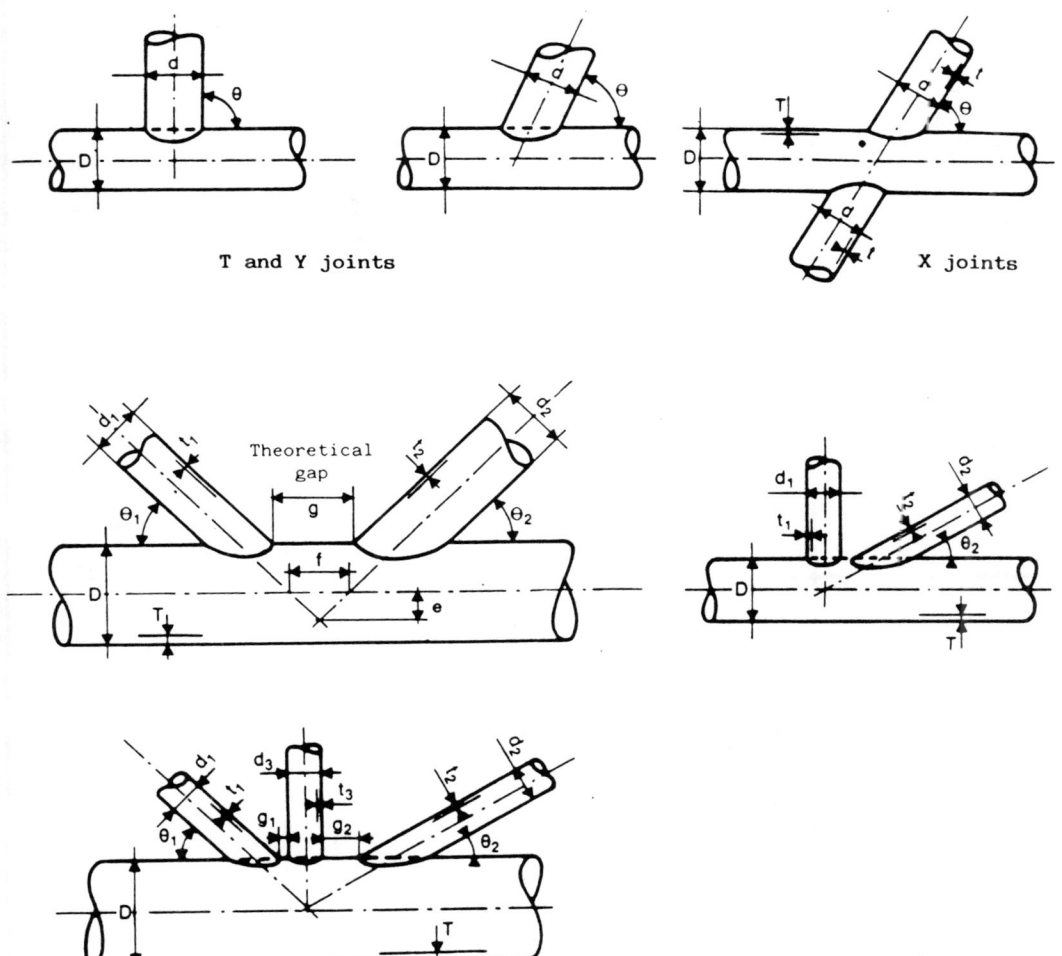

T and Y joints                          X joints

N, K and KT joints

| Load—Geometry | SCF | Validity |
|---|---|---|
| **T, Y** Axial load | CHORD<br><br>$SCF = 1.981.\alpha^{0,057}.e^{-1.2\beta^3}$<br>$.\gamma^{0,808}.\tau^{1.333}.\sin^{1,694}\theta$ | |
| | BRACE<br><br>$SCF = 3.751.\alpha^{0.12}.e^{-1.35\beta^3}$<br>$.\gamma^{0.55}.\tau.\sin^{1.94}\theta$ | |
| **T, Y** In-plane bending load | CHORD<br><br>$SCF = 0.702.\beta^{-0.04}$<br>$.\gamma^{0.6}.\tau^{0.86}.\sin^{0.57}\theta$ | $6.6 \leq \alpha \leq 40$<br>$0.3 \leq \beta \leq 0.8$<br>$8.3 \leq \gamma \leq 33.3$<br>$0.2 \leq \tau \leq 0.8$<br>$0,01 \leq \zeta \leq 1.0$<br>$0° \leq \theta° \leq 90°$ |
| | BRACE<br><br>$SCF = 1.301.\beta^{-0.38}$<br>$.\gamma^{0.23}.\tau^{0.38}.\sin^{0.21}\theta$ | |
| **T, Y** Out-of-plane bending load | CHORD $\quad 0.3 \leq \beta \leq 0.55$<br>$SCF = 1.024.\beta^{0.787}$<br>$.\gamma^{1.014}.\tau^{0,889}.\sin^{1.557}\theta$ | |
| | BRACE $\quad 0.3 \leq \beta \leq 0.55$<br>$SCF = 1.522.\beta^{0,801}$<br>$.\gamma^{0.852}.\tau^{0.543}.\sin^{2.033}\theta$ | |
| | CHORD $\quad 0.55 \leq \beta \leq 0.75$<br>$SCF = 0.462.\beta^{-0.619}$<br>$.\gamma^{1.014}.\tau^{0.889}.\sin^{1.557}\theta$ | |
| | BRACE $\quad 0.55 \leq \beta \leq 0.75$<br>$SCF = 0.796.\beta^{-0.281}$<br>$.\gamma^{0.852}.\tau^{0.543}.\sin^{2.033}\theta$ | |

**EPR FORMULAS**

| Load-Geometry | SCF | Validity |
|---|---|---|
| **.K,N**<br><br>Axial load | **CHORD**<br><br>$SCF = 1.506.\beta^{-0.059}.\gamma^{0.666}$<br>$.\tau^{1.104}.\zeta^{0.067}.\sin^{1.521}\theta$<br><br>**BRACE**<br><br>$SCF = 0.920.\beta^{-0.441}.\gamma^{0.157}$<br>$.\tau^{0.560}.\zeta^{0.058}.e^{1.448 \sin\theta}$ | |
| **K,N**<br><br>In-plane bending load | **CHORD**<br><br>$SCF = 1.822.\beta^{0.06}.\gamma^{0.38}$<br>$.\tau^{0.94}.\sin^{0.9}\theta$<br><br>**BRACE**<br><br>$SCF = 2.827.\beta^{-0.35}$<br>$.\tau^{0.35}.\sin^{0.5}\theta$ | $6.6 \le \alpha \le 40$<br>$0.3 \le \beta \le 0.8$<br>$8.3 \le \gamma \le 33.3$<br>$0.2 \le \tau \le 0.8$<br>$0.01 \le \zeta \le 1.0$<br>$0° \le \theta° \le 90°$ |
| **KT**<br><br>Axial load | **CHORD**<br><br>$SCF = 1.832.\beta^{0.12}.\gamma^{0.10}$<br>$.\tau^{0.68}.(\zeta_1+\zeta_2)^{0.126}.\sin^{0.5}\theta$<br><br>**BRACE**     $0° \le \theta° \le 45°$<br>$SCF = 6.056.\beta^{-0.36}.\gamma^{0.10}$<br>$.\tau^{0.68}.(\zeta_1+\zeta_2)^{0.126}.\sin^{0.5}\theta$<br><br>**BRACE**     $45° \le \theta° \le 90°$<br>$SCF = 13.804.\beta^{-0.36}.\gamma^{0.10}$<br>$.\tau^{0.68}.(\zeta_1+\zeta_2)^{0.126}.\sin^{2.88}\theta$<br><br>**CENTRAL BRACE**     $0° \le \theta_2° \le 90°$<br>$SCF = 4.891.\beta^{-0.396}.\gamma^{0.123}$<br>$.\tau^{0.672}.(\zeta_1+\zeta_2)^{0.159}.\sin^{2.267}\theta_2$ | $\theta_2$: inclination of central brace |

**EPR FORMULAS (continued)**

| Load-Geometry | SCF | Validity |
|---|---|---|
| **T, Y**<br><br>Axial load | CHORD<br><br>$$SCF = \left(1.44 - 3.72 \cdot (\beta - 0.47)^2\right) \cdot \gamma^{0.87}$$<br>$$\cdot \tau^{1.37} \cdot \alpha^{0.06} \cdot \sin^{1.694}\theta$$<br><br>BRACE<br><br>$$SCF = \left(1.00 - 1.78 \cdot (\beta - 0.5)^2\right) \cdot \gamma^{0.76}$$<br>$$\cdot \tau^{0.57} \cdot \alpha^{0.12} \cdot \sin^{1.94}\theta$$ | CHORD<br><br>$7.0 \leq \alpha \leq 40$<br>$0.255 \leq \beta \leq 0.9$<br>$10 \leq \gamma \leq 30$<br>$0.4 \leq \tau \leq 1.0$<br>$0° \leq \theta° \leq 90°$ |
| **T, Y**<br><br>In-plane bending load | CHORD<br><br>$$SCF = \left(1.65 - 1.1 \cdot (\beta - 0.42)^2\right) \cdot \gamma^{0.38}$$<br>$$\cdot \tau^{1.05} \cdot \sin^{0.57}\theta$$<br><br>BRACE<br><br>$$SCF = \left(0.95 - 0.65(\beta - 0.41)^2\right) \cdot \gamma^{0.39}$$<br>$$\cdot \tau^{0.29} \cdot \sin^{0.21}\theta$$ | BRACE<br><br>$7.0 \leq \alpha \leq 16.0$<br>$0.3 \leq \beta \leq 0.9$<br>$10.0 \leq \gamma \leq 30.0$<br>$0.47 \leq \tau \leq 1.0$<br>$0° \leq \theta° \leq 90°$ |
| **T, Y**<br><br>Out-of-plane bending load | CHORD<br><br>$$SCF = \left(1.01 - 3.36 \cdot (\beta - 0.64)^2\right) \cdot \gamma^{0.95}$$<br>$$\cdot \tau^{1.18} \cdot \sin^{1.557}\theta$$<br><br>BRACE<br><br>$$SCF = \left(0.76 - 1.92 \cdot (\beta - 0.72)^2\right) \cdot \gamma^{0.89}$$<br>$$\cdot \tau^{0.47} \cdot \sin^{2.033}\theta$$ | Term $\alpha$ of EPR added by DnV<br><br>Term $\sin\theta$ of EPR added by CTICM |

**DnV FORMULAS**

| Load-Geometry | SCF | Validity |
|---|---|---|
| **T, Y** Axial load | **CHORD** $$SCF_{SP} = \beta.(6.78-6.42.\beta^{0.5})$$ $$.\gamma.\tau.\sin^{(1.7+0.7\beta^3)}\theta$$ $$SCF_{CP} = k'_c + k_o k''_c$$ | $8 < \alpha < 40$ $0.13 \leq \beta \leq 1.0$ $12 < \gamma < 32$ $0.25 < \tau < 1.0$ $30° < \theta < 90°$ $\underline{SCF}_{SP} = SCF$ at saddle point $\underline{SCF}_{CP} = SCF$ at crown point |
| | **BRACE** $$SCF_{SP} = 1.0 + 0.63. \, SCF_{SP.CHORD}$$ $$SCF_{CP} = 1.0 + 0.63. \, SCF_{CP.CHORD}$$ | |
| | $k'_c = (0.7+1.37.(1-\beta).\gamma^{0.5}.\tau).(2.\sin^{0.5}\theta-\sin^3\theta)$ $k_o = \tau.(\beta-\tau.(2.\gamma)^{-1}).(0.5.\alpha-\beta.\sin^{-1}\theta).\sin\theta.(1-1.5\gamma^{-1})^{-1}$ $k''_c = 1.05+(30,0.\gamma^{-1.0}.\tau^{1.5}.(1.2-\beta).(\cos^4\theta+0.15))$ | |
| **T, Y      K,N** **X**          **KT** In-plane bending load | **CHORD** $$SCF_{CP} = 0.75.\gamma^{0.60}.\tau^{0.8}$$ $$.(1.6.\beta^{0.25}-0.7.\beta^2).\sin^{(1.5-1.6\beta)}\theta$$ **BRACE** $$SCF_{CP} = 1.0 + 0.63. \, SCF_{SP.CHORD}$$ | $8 < \alpha < 40$ $0.13 \leq \beta \leq 1.0$ $12 < \gamma < 32$ $0.25 < \tau < 1.0$ $30° < \theta < 90°$ $0.0 \quad \leq \zeta$ |
| **T, Y** Out-of-plane bending load | **CHORD** $$SCF_{SP} = \beta.(1.6-1.15\beta^5).\gamma.\tau$$ $$.\sin^{(1.35+\beta^2)}\theta$$ **BRACE** $$SCF_{SP} = 1.0 + 0.63. \, SCF_{SP.CHORD}$$ | $\underline{SCF}_{SP} = SCF$ at saddle point $\underline{SCF}_{CP} = SCF$ at crown point |

**LLOYD'S FORMULAS**

| Load-Geometry | SCF | Validity |
|---|---|---|
| **X**<br><br>Axial load | **CHORD**<br>$$SCF_{SP} = 1.7 \cdot \gamma \cdot \tau \cdot \beta \cdot (2.42 - 2.28 \cdot \beta^{2.2})$$<br>$$\cdot \sin^{\beta^2} \cdot (15 - 14.4\beta)_\theta$$<br><br>**BRACE**<br>$$SCF_{SP} = 1.0 + 0.63 \, {}^{SCF}SP.CHORD$$ | |
| **X**<br><br>Out-of-plane bending load | **CHORD**<br>$$SCF_{SP} = \beta \cdot (1.56 - 1.46\beta^5) \cdot \gamma \cdot \tau$$<br>$$\cdot \sin^{\beta^2} \cdot (15 - 14.4\beta)_\theta$$<br><br>**BRACE**<br>$$SCF_{SP} = 1.0 + 0.63 \, {}^{SCF}SP.CHORD$$ | $8 < \alpha < 40$<br>$0.13 \leq \beta \leq 1.0$<br>$12 < \gamma < 32$<br>$0.25 < \tau < 1.0$<br>$30° < \theta < 90°$<br>$0.0 \leq \zeta$<br><br><br>$\underline{SCF}_{SP} = SCF$<br>at saddle point<br><br>$\underline{SCF}_{CP} = SCF$<br>at crown point |
| **KT**<br><br>Out-of-plane bending load | **CENTRAL CHORD**    $M_1 = M_2 = M_3$<br>$$\theta_2 > \theta_1 \quad \theta_2 > \theta_3 \quad \theta_1 = \theta_3$$<br>$$SCF_{SP} = \Big( SCF_{SP} \text{ y JOINT } \theta = \theta_2 \Big)$$<br>$$\cdot \Big( 1.0 + 2.0(\theta_2/\theta_1) \Big)^{0.3}$$<br>$$\cdot (\sin\theta_1 / \sin\theta_2)^{(1.35+\beta^2)}$$<br>$$\cdot (0.016 \cdot \gamma \cdot \beta)^{(\zeta+0.45)} \Big)$$<br>$$\cdot (1.0 - 0.1^{(1.0+4\zeta)})^2$$<br><br>**CENTRAL BRACE**<br>$$SCF_{SP} = 1.0 + 0.63 \cdot {}^{SCF}SP.CHORD$$ | $\theta_2$: inclination of central brace |

**LLOYD'S FORMULAS** (continued)

| Load–Geometry | SCF | Validity |
|---|---|---|
| **K,N** **KT** Axial load | CHORD $$\theta_1 \geq \theta_2 \quad ; \quad P_2 = P_1 \cdot \sin\theta_1 / \sin\theta_2$$ $$SCF_{SP} = \left( \begin{array}{c} SCF_{SP} \\ y \text{ JOINT } \theta = \theta_1 \end{array} \right)$$ $$\cdot \left( 1.0 - (0.012 \cdot \gamma)^{(0.67 \cdot \zeta + 0.4)} \right.$$ $$\left. \cdot (\sin\theta_1 / \sin\theta_2)^{(0.1 - 0.7\beta^3)} \right)$$ $$SCF_{CP} = 1.1 \cdot \gamma^{0.65} \cdot \tau \cdot \sin\theta_1 \cdot \sin\theta_2^{-0.5}$$ $$\cdot (2.0 \cdot \zeta)^{0.05\beta} \cdot (1.5\beta^{0.25} - \beta^2)$$ ——————————————— BRACE $$SCF_{SP} = 1.0 + 0.63 \cdot SCF_{SP.CHORD}$$ $$SCF_{CP} = 1.0 + 0.63 \cdot SCF_{CP.CHORD}$$ | $8 < \alpha < 40$ $0.13 \leq \beta \leq 1.0$ $12 < \gamma < 32$ $0.25 < \tau < 1.0$ $30° < \vartheta < 90°$ $0 \quad \leq \zeta$ |
| **K,N** Out-of-plane bending load | CHORD $\qquad \theta_1 \geq \theta_2 \qquad M_1 = M_2$ $$SCF_{SP} = \left( \begin{array}{c} SCF_{SP} \\ y.JOINT\theta = \theta_1 \end{array} \right)$$ $$\cdot \left( 1.0 + (0.016 \cdot \gamma \cdot \beta)^{(\zeta + 0.45)} \right.$$ $$\cdot (\theta_1/\theta_2)^{0.33} \cdot (\sin\theta_2/\sin\theta_1)^{(1.35+\beta^2)} )$$ $$\cdot (1.0 - 0.1^{(1.0+4\zeta)})$$ ——————————————— BRACE $$SCF_{SP} = 1.0 + 0.63 \cdot SCF_{SP.CHORD}$$ | $\underline{SCF}_{SP} = SCF$ at saddle point $\underline{SCF}_{CP} = SCF$ at crown point |

**LLOYD'S FORMULAS (end)**

ACHEVÉ D'IMPRIMER
SUR LES PRESSES DE
L'IMPRIMERIE CHIRAT
42540 ST-JUST-LA-PENDUE
EN OCTOBRE 1987
DÉPÔT LÉGAL 1987 N° 3441
N° D'ÉDITEUR 752